GUIDELINES FOR

Postrelease Mitigation Technology in the Chemical Process Industry

This *Guidelines* is one of a series of publications available from the Center for Chemical Process Safety. A complete list of titles appears at the end of this book.

Postrelease Mitigation Technology in the Chemical Process Industry

CENTER FOR CHEMICAL PROCESS SAFETY

of the

American Institute of Chemical Engineers
345 East 47th Street, New York, NY 10017

Library of Congress Cataloging-in Publication Data
Guidelines for postrelease mitigation technology in the chemical
  process industry
        p.   cm.
  Includes bibliographical references (p.   –   ) and index.
  ISBN 0–8169–0588–6
  1. Hazardous substances—Safety measures.   2. Hazardous
substances—Environmental aspects.   3. Chemical industries—Safety
measures   I. American Institute of Chemical Engineers. Center
for Chemical Process Safety.
T55.3.H3G8527   1997
660′ .2804—dc21   96–45242
                                              CIP

Contents

Preface

For 40 years, the American Institute of Chemical Engineers has been involved with process safety and loss control issues in the chemical, petrochemical, hydrocarbon process, and related industries and facilities.

The Center for Chemical Process Safety (CCPS), a directorate of AIChE, was established in 1985 to develop and disseminate information for use in the prevention of major chemical incidents. With the support and direction of the CCPS Advisory Managing Boards, a multifaceted program was established to address the need for process safety management systems to reduce potential exposures to the public, facilities, personnel, and the environment This program involves the development and publication of guidelines relating to specific areas of process safety management; organizing, convening and conducting seminars, symposia, training programs, and meetings on process safety-related matters; and cooperation with other organizations, both internationally and domestically, to promote process safety. CCPS's activities are supported by funding and professional expertise by over 90 entities.

In 1988, CCPS published *Guidelines for Vapor Release Mitigation*, a survey of then current industrial practice for controlling accidental releases of hazardous vapors and preventing their escape from the source area. Its focus was primarily on prerelease factors, including inherently safer plants and engineering design matters. As the title states, the present book is concerned primarily with postrelease factors in mitigation.

Acknowledgments

The American Institute of Chemical Engineers and the Center for Chemical Process Safety (CCPS) express their gratitude to all the members of the Vapor Cloud Modeling Subcommittee for their unstinting efforts and technical contributions in the preparation of this Guidelines. The members of this distinguished group are:

Ronald J. Lantzy, *Chair*	(Rohm and Haas Company)
Gib R. Jersey, *Vice Chair*	(Mobil Technology Company)
William J. Hague, *past Chair*	(Allied Signal Inc.)
Douglas N. Blewitt	(Amoco Corporation)
Sanford G. Bloom	(Lockheed Martin Energy Systems)
Donald J. Connolley	(AKZO Nobel Chemicals, Inc.)
George E. DeVaull	(Shell Oil Company)
Albert G. Deitz, Jr.	(U.S. Department of Energy)
Seshu Dharmavaram	(DuPont Company)
David J. Fontaine	(Chevron Research and Technology Co.)
Gene K. Lee	(Air Products & Chemicals, Inc.)
David McCready	(Union Carbide Corporation)
John T. Marshall	(Dow USA)
Robert L. Moser	(Cigna Insurance Company)
Malcolm L. Preston	(ICI Engineering Technology)
Gerry M. Schray	(Monsanto Company)
Kenneth W. Steinberg	(Exxon Research and Engineering Co.)
Jawad Touma	(U.S. EPA)

CCPS guidance and counsel were appropriately provided by its Directors, Thomas W. Carmody (retired) and Bob G. Perry. Liaison between the subcommittee and CCPS was provided by William J. Minges, CCPS Staff.

The contract for preparing the Guidelines was awarded to Arthur D. Little, Cambridge, MA. Each of the following from that organization was a significant contributor to the publication:

Frederick T. Dyke

Kumkum M. Dilwali

Marian H. Long

Georges A. Melhem

Steven R. Radis

R. Peter Stickles

In addition, invaluable secretarial assistance was provided by Pat Tryon Cutting.

CCPS also expresses its appreciation to members of the Technical Steering Committee for their valuable advice and support. The group to whom the Subcommittee is especially indebted consists of those who volunteered to provide peer review:

Myron L. Casada (JBF Associates, Inc.)
Daniel A. Crowl (Michigan Technological University)
Jim Evans (Union Carbide Corporation)
Rudolph Frey (The M.W. Kellogg Company)
Thomas O. Gibson (The Dow Chemical Company)
John A. Hoffineister (Lockheed Martin Energy Systems)
Greg Hounsell (Pfizer, Inc.)
Gregory Keeports (Rohm and Haas Company)
Peter N. Lodal (Eastman Chemical Company)
John A. Noronha (Eastman Kodak Company)
James L. Paul (Hoechst Celanese Corporation)
Frank P. Ragonese (Mobil Oil Corporation)
Lester H. Wittenberg (AIChE/CCPS)

1

Introduction to Postrelease Mitigation

1.1. Introduction

The safe operation of chemical facilities and protection of the people and the environment surrounding them is a requirement for creating long-term social and economic benefits for the community and industry. Many of the major accidents in the chemical process industry have involved the release of toxic vapors or ignition of flammable vapors. The purpose of this book is to address current engineering methods in practice for minimizing the consequences of these types of releases, once the event has occurred.

Techniques for prevention and control of accidental chemical releases can be broadly classified into methods that eliminate the causes of a release, reduce the probability of a release, or reduce the impact of a release. By far, the most beneficial methods for improved process safety focus on avoiding accidents and inadvertent releases. These methods include administrative controls such as procedures, training, maintenance, and inspections. Other successful preventative measures may focus on engineering design or redesign, process changes, improved process control, minimizing chemical inventory, or engineering countermeasures. *Guidelines for Technical Management of Chemical Process Safety* (CCPS, 1989) and *Guidelines for Safe Storage and Handling of High Toxic Materials* (CCPS, 1988b) describe these preventative methods.

The techniques that can be applied as postrelease mitigation measures in many cases are not based on mature science. Only in recent years has attention been focused on the design of postrelease mitigation systems and the development of needed fundamental knowledge through testing and data correlation.

Empirical data are available for the design of some postrelease mitigation systems but these cases are limited in scope and only applicable to specific situations.

The use of these data to design similar systems for other compounds or situations may require an extrapolation beyond the data's range. In these situations the use of the data requires careful engineering judgment.

Fundamental theoretical equations that describe a phenomenon can be written for many situations. However, the range over which such equations might be useful and what corrections might have to be applied to account for nonideal situations are unknown because little data exist to validate the equations. At the same time, these equations do provide an insight into what important parameters should be tested.

CCPS (1988a) has published *Guidelines for Vapor Release Mitigation*, which covers many aspects of current industrial practice in controlling accidental releases of hazardous vapors. Since the publication of *Guidelines for Vapor Release Mitigation* (and partly because of its publication), great progress and improvements have been made in many areas of mitigation design. At present, most of this newer material is available only in technical reports, journals, and scientific proceedings. The purpose of this book is to collect and present recent material in a guideline format. Many of the topics discussed in *Guidelines for Vapor Release Mitigation* remain timely; however, this book focuses more on practical engineering design of mitigation systems and in particular, on postrelease mitigation methods. In addition, certain mitigation techniques, such as refrigeration, which would normally be classified as prerelease mitigation, have been included here.

CCPS also realizes that regarding some of these mitigation systems our current understanding of the fundamental physical processes is rudimentary and further research and development are needed.

1.2. Scope of This Book

The principal focus of this book is the mitigation of accidental releases of toxic or flammable materials through release countermeasures, in particular, postrelease systems. Postrelease systems are designed for control of a hazardous material once it has been released into the environment. Control measures can include passive systems, such as dikes or berms around storage tanks, as well as active methods, such as water-spray or deluge systems installation around a process unit, or application of foam on a chemical spill. However, fire fighting, blast protection and environmental control of response methodologies are not covered in this guideline.

1.3. Benefits of Postrelease Mitigation Techniques

The management of process safety is many-faceted. The mitigation systems discussed in this book are only part of a total solution. Structured evaluation of chemical process safety is a process of continuous improvement that includes problem identification, development of proposed solutions, evaluation of solutions against established criteria, and, if appropriate, implementation of solutions. This approach is shown in Figure 1.1.

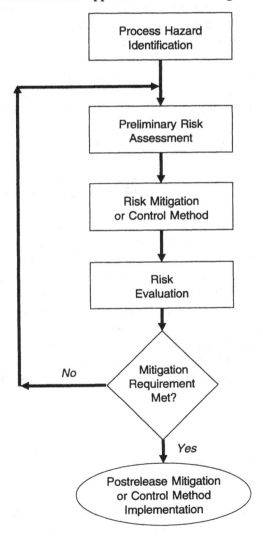

Figure 1.1. Implementation process for mitigation evaluation. [From (CCPS, 1993).]

The potential hazards of chemical vapor releases, including toxicity, are primarily related to concentration levels and time of exposure at specific locations, as discussed in Chapter 3. Mitigation methods discussed in this book are intended to reduce or eliminate significant acute (short-term) effects of a dispersing vapor cloud in regions of concern. Such regions of concern could include the facility itself, the neighboring industries, and the surrounding community, with different evaluation criteria for each area. Within an industrial environment, the basis of these criteria will include the ability to control site access, and to inform, train and equip every individual for safe work practices and emergency response procedures. In the general community the exposure criteria are generally more stringent, and must include the possibility of exposure of sensitive individuals.

Reducing the potential impacts of a vapor release means reducing either the concentration levels or time of exposure, or both, in the regions of concern. Within the context of postrelease mitigation of vapor or aerosol/vapor clouds, this means improved dilution by reducing the amount of material going into the air, removing a fraction of the chemical of concern from a cloud, and/or increasing the entrainment of ambient air into a cloud. A well-designed water spray (with a compatible chemical) is one potential method for achieving both removal and dilution in a given vapor release scenario.

With a volatile liquid spill there are broader options available for postrelease mitigation of the evolved vapor. Reducing the spill volatilization rate alone, with foam or dilution of the spilled chemical pool, would decrease the downwind vapor concentrations, but could also increase the duration of the vapor release, since the liquid pool would be present for a longer period of time. However, reducing the volatilization rate also provides the opportunity for a second option, which is source control, with removal, recovery or neutralization of the remaining liquid pool. Prior planning also applies here; for example, the volatilization rate of a chemical spill on a sun-warmed concrete pad can be reduced by providing shade for the pad and preventing it from getting hot initially.

1.4. How to Use This Guideline

The purpose of this guideline is to present methods and design examples of postrelease mitigation methods which are in current industrial use, and to assess areas where further research is needed.

Because of the wide range of potential release scenarios which can be encountered, the techniques applied for control of accidental releases necessarily require good engineering judgment. The information in this guideline is not intended to define when to apply postrelease mitigation methods. Such criteria are better managed within the context of process safety management (CCPS, 1989), with postrelease mitigation being one option among many that are available for the control of accidental hazardous chemical releases.

1.5. Guideline Organization and Content

This guideline has been organized to meet the needs of readers who are new to the concept of postrelease mitigation and who are interested in learning about techniques available for mitigating specific scenarios. To help each reader determine where to look in the guideline, a short description of each chapter's contents is provided below.

Chapter 2—Overview of Release Scenarios and Postrelease Mitigation
This chapter provides an overview of the different types of releases that can occur, the consequences of the discharges, and a general description of available postrelease mitigation measures.

Chapter 3—Vaporization Reduction
This chapter describes the mechanism of vaporization from a liquid pool and the techniques that can be used to reduce the rate of vaporization of a spilled liquid.

Chapter 4—Fluid Curtains
This chapter covers the use of fluid curtains of water, steam, and air to mitigate the effects of a release.

Chapter 5—Secondary Containment
This chapter discusses various containment techniques that can be utilized to control a released material.

Chapter 6—Detection and Response
This chapter provides an overview of means to detect a release that has occurred and support planning for emergency response.

Chapter 7—Examples of Mitigation Effectiveness
This chapter uses consequence modeling to evaluate the effects of various postrelease mitigation measures.

1.6. References

CCPS (Center for Chemical Process Safety) 1988a. *Guidelines for Vapor Release Mitigation.* New York: American Institute of Chemical Engineers.
CCPS (Center for Chemical Process Safety) 1988b. *Guidelines for Safe Storage and Handling of High Toxic Materials.* New York: American Institute of Chemical Engineers.
CCPS (Center for Chemical Process Safety) 1989. *Guidelines for Technical Management of Chemical Process Safety.* New York: American Institute of Chemical Engineers.
CCPS (Center for Chemical Process Safety) 1993. *Guidelines for Safe Automation of Chemical Processes.* New York: American Institute of Chemical Engineers.

2

Overview of Release Scenarios and Postrelease Mitigation

2.1. Introduction

This chapter introduces the various methods of postrelease mitigation. First, however, prerelease mitigation techniques, different types of releases, and the potential consequences of hazardous material releases are discussed. These discussions are intended to provide the reader with some background into postmitigation concepts covered in detail later. An experienced engineer can use this chapter as a quick reference to the release mitigation techniques available. For someone new to this subject, this chapter will help to focus on appropriate strategies which are explored more fully in later chapters.

For a more detailed technical discussion of release mechanisms and their consequences, refer to other CCPS publications (CCPS, 1988a,b, 1989a, 1996).

Before any mitigation measures can be designed, an effective hazard identification study must be conducted. The results of such a study (a set of release scenarios) can be used to develop a coherent set of mitigation strategies. In the process industries, these studies are most commonly conducted using hazard and operability (HAZOP) studies, what-if checklists, failure modes and effects analyses (FMEA), and several other comparable techniques (CCPS, 1992).

2.2. Mitigation Categories

Mitigation techniques are divided into two categories: prerelease and postrelease mitigation measures. Prerelease mitigation measures take effect before any material is released to the environment. Prerelease measures minimize the likelihood of a release, or destroy the material before it is released, or modify the material's properties to reduce the extent of any hazard following release. postrelease mitigation measures reduce the consequences of a release once it has reached the environment.

Flare stacks that safely burn organic material released into vent headers are an example of a prerelease mitigation measure. The flare destroys the hazardous organic material before it reaches the environment. A dike around a storage tank is an example of a postrelease mitigation measure. The dike contains the release in a small area, reducing the total evaporation rate from the spill and so reducing the impact of the release.

2.3. Prerelease Mitigation Techniques

Prerelease mitigation techniques are designed to prevent or minimize the uncontrolled release of hazardous materials. Detailed discussion of these techniques can be found in the Center for Chemical Process Safety books, *Guidelines for Vapor Release Mitigation* (CCPS, 1988a, Section 3.0), and *Guidelines for Safe Storage and Handling of High Toxic Hazard Materials* (CCPS, 1988b). The main approaches and some illustrative examples are briefly described below to provide an overview of prerelease mitigation.

2.3.1. Inherently Safer Design

An inherently safer design eliminates or minimizes hazards by changing process conditions or process chemistry, and by selecting appropriate locations for hazardous operations. Inherently safer design includes the following approaches:

- Reduce inventory such that the impacts associated with a potential release are reduced to an acceptable level.
- Substitute harmless or less hazardous materials for those that are more hazardous.
- Modify the process to reduce the consequences of a release; for example, use refrigerated rather than pressurized storage, which results in higher liquid release and lower vapor release rates (from the same hole size) and lower vapor generation rates from a pool; dilute hazardous materials with inert or less hazardous material to reduce partial pressure; or reduce operating pressures to reduce release rates.

- Site new facilities away from population centers or other hazardous facilities and employ large buffer zones around facilities to prevent large populations growing in the immediate vicinity of the hazards. To be effective, these buffer zones should be owned, or at least controlled, by the operator to prevent encroachment of populations.

2.3.2. Physical Integrity of a Plant

Releases occur because the physical envelope (consisting of vessels, pipework, valves, tanks, etc.) that contains the process fluids has been breached. A breach can occur because a drain or vent valve has opened or a relief device has operated; such breaches can be attributed to an operating error or the correct functioning of a safety system. The physical containment can also be breached by the mechanical failure of a piece of equipment; for example, the failure of a pump seal, a flange leak, or the erosion of a pipe bend or corrosion of piping. Retaining the physical integrity of a plant requires good mechanical design, proper selection of materials of construction as well as thorough scheduled inspection and maintenance.

The key elements of a process safety management system used to assure physical integrity are described in two books, *Guidelines for Technical Management of Chemical Process Safety* (CCPS, 1989b) and *Guidelines for Safe Storage and Handling of High Toxic Hazard Materials* (CCPS, 1988b). The following topics are presented in these references:

- Good design practices to ensure that the mechanical design of the equipment is suited to normal operation, startup, shutdown, and emergency conditions (for example, ensuring that vessel wall thickness, design temperature and pressure, corrosion allowance, etc., adhere to industry or national codes and standards);
- Selection of appropriate materials of construction that are resistant to erosion, corrosion, or extreme temperature will preclude a loss of physical integrity;
- Management-of-change procedures that require proper review of all changes to the design or operation of a plant;
- Pre-startup safety reviews to verify that all equipment has been properly installed, that operating, maintenance, and emergency procedures are in place, and that all staff have been adequately trained;
- Effective inspection and testing during construction and operations, which are essential to ensure that equipment was constructed as designed, was properly installed, and is working correctly.

2.3.3. Process Integrity

Process integrity addresses the reactions, physical chemistry, and dynamics of a plant's operation. If these are allowed to run out of control or exceed standards, process conditions may exceed the mechanical design limits of the equipment. Some of the following approaches and methods used to assure process integrity are described in *Guidelines for Technical Management of Chemical Process Safety* (CCPS, 1989b):

- Comprehensively identify all reactants, solvents and intermediates to minimize the inadvertent mixing of incompatible chemicals (proper labeling/segregation).
- Specify limits on operating conditions (critical operating parameters) that, if exceeded, could result in hazardous conditions (critical operating limits) *which then trip an alarm and/or automatic shutdown systems.* These critical operating parameters are set below the point at which hazardous conditions might occur.
- Establish process control systems capable of maintaining conditions within safe limits (may include emergency cooling or heating systems).
- Install, inspect, and maintain properly sized pressure-relief systems to prevent overpressurization of equipment.

2.3.4. Emergency Relief Treatment Systems

Occasionally process conditions exceed normal operating limits because of equipment or operating malfunction, and the plant safety systems vent the process materials to relieve any overpressure. If the process material is hazardous, an emergency relief disposal system could be provided. The CCPS book entitled *Guidelines for Vapor Release Mitigation* (CCPS, 1988a) describes the following emergency relief treatment systems in some detail:

- *Active scrubbers*, which include pumps or other active components. These scrubbers generally circulate chemicals (such as water, acid, caustics, or organics) through a packed or trayed tower to absorb and/or condense hazardous materials from a vapor stream. The active components may run continuously, or only when material is released to the scrubber. This approach is most effective when the circulating fluid reacts with the material being adsorbed or condensed.
- *Elevated stacks* to aid atmospheric dispersion of the vapor releases.
- *Flares* to burn any hazardous organic materials in the streams.
- *Catchtanks* to hold vented material and to separate vapor and liquid streams prior to safe disposal.

- *Quench tanks* filled with liquid to cool and condense vapor releases, neutralize reactive materials, or cool and quench reaction masses. In these systems, a large volume of quench material is needed to deal with the heat effects because they are static passive scrubbers with no heat exchange capabilities.

2.3.5. Emergency Process Abort Systems

Another technique for mitigating process upsets is to abort the process by means of an emergency process control system. This approach is most commonly employed with reactive chemical systems. Successful methods include:

- Emergency cooling systems to slow reaction and/or reduce pressure buildup;
- Dump systems to divert reactants to a safer location, such as catch-tanks or quench tanks;
- Reaction quench or chemical "short stop" systems to stop or slow the reaction by removing heat sources, cooling the reactants, deactivating any catalyst, or interfering with reaction chemistry;
- Manual or automatic emergency shutdown systems to shut off one or more reactant feeds and then vent the contents to reduce reactor pressure;
- Emergency heating systems when a material could either freeze or become viscous, which would plug lines and prevent movement into a mitigation system.

2.3.6. Emergency Isolation of Releases

In many instances the point of release can be isolated from the inventory of hazardous material feeding the release; this way the quantity of material released can be minimized. The emergency isolation-of-releases method requires that isolation valves be installed between possible release points and any large inventory of hazardous material. Release points of concern include pump seals, small pipe fittings, and compressors. This approach is addressed in the book entitled *Guidelines for Vapor Release Mitigation* (CCPS, 1988a) under the following topics:

- Isolation devices, such as manual/automatic valves, check/excess-flow valves, "dry-break" couplings, which shut off flow if the coupling is parted, and positive displacement pumps, which stop flow when shut down;

- Automatic valves operated from safe, remote locations either through interlock systems or switches to isolate hazardous inventories from the point of release.

These devices should be designed and installed so that they can be routinely inspected and tested without shutting down the equipment they are protecting. Proper inspection and testing of these devices is intended to assure their reliability.

To be effective, the inventory must be isolated quickly to prevent its total contents from being released. Since total release may occur within a few minutes, rapid response and operation is important.

2.4. Release Scenarios and Consequences

To quantify the consequences of hazardous events, their flow rate, duration and phase (gas, liquid, solid) must be determined accurately. Flow rate, duration and phase are functions of storage and release conditions, the properties of the chemical or mixture of chemicals, and the nature of the containment failure.

This guideline covers only nonroutine or accidental events. Many hazardous events start with the discharge or loss of containment of a flammable and/or toxic material from a vessel or pipe. These discharges, which may take the form of vapor, liquid, solid, or multiphase vapor–liquid–solid mixtures, may be released into a confined area, such as a dike, building, or an equipment array, or into an open, unconfined area. The sources of these releases could be holes in vessels or pipelines, open pressure-relief devices, pipe ruptures, flange and seal leaks, or catastrophic vessel ruptures. The range of releases is illustrated in Figure 2.1.

2.4.1. Types of Releases

To determine the consequences of a release, it is first necessary to calculate the mass flow rate or quantity of material released and the state of the fluid being released. For "design" releases, for example, from safety relief devices or drain valves, the analysis is generally a straightforward engineering exercise. In the case of a mechanical failure, the analysis is complicated by the unknown characteristic of the hole, including its cross-sectional area and shape. For example, a pipe may be snapped off by a falling object, could split at a weld, or open at an erosion point. The resulting holes would have very different flow rate characteristics.

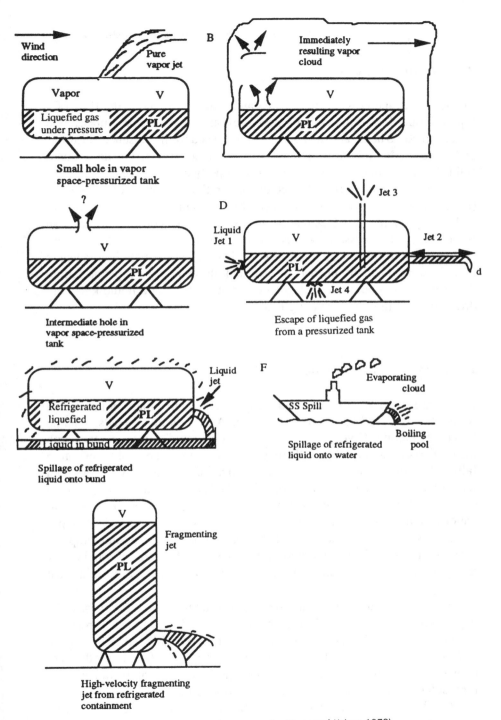

Figure 2.1. Some release scenarios (Fryer and Kaiser, 1979).

2.4.2. Liquid Releases

In addition to the cross-sectional area and shape of the hole, the flow of a liquid is controlled by the available pressure (vapor space pad pressures, pump head, and liquid head) and the pressure drop caused by the presence of fittings and valves in the fluid flow path. Flow rates can be estimated using standard fluid flow equations when the opening is properly characterized.

A liquid release offers the potential for the formation of an aerosol (mist). Aerosol formation is influenced by mechanical force (acceleration) and fluid properties, such as the material's surface tension. Additionally, in some systems the pressure, because of static head, may be adequate to cause some aerosol formation and must be considered. Examples of liquid releases include an accidentally opened storage tank drain valve, a pipeline failure downstream of a pump, a damaged nozzle at the base of a liquid knockout drum, or the failure of a loading hose.

2.4.3. Liquid Pool Formation

In the case of a nonpressurized liquid spill that occurs either on land or on water, the spill spreads over the surface. As the liquid spreads, volatile liquids evaporate. Cryogenic liquids often boil rapidly upon contact with the ground or water until the substrate is cooled and/or the pool temperature falls below the boiling point. While vaporization continues when the pool temperature falls below the boiling point, the first few minutes following a cryogenic liquid spill are usually the most significant with respect to vaporization. Evaporation rates of spilled liquids depend upon their vapor pressure, storage temperatures, meteorological conditions, ground temperature, and the like. The evaporation rate is a strong function of ambient air and ground temperature and wind velocity. It is also affected by solar heat flux.

For liquids with low boiling points, such as most liquefied gases, the vaporization rate is normally driven by the rate of heat transfer from the ground by conduction and, to a lesser extent, by wind speed. Evaporation and boiling rates vary with time as the pool and the surface beneath it are cooled. The ground temperature is strongly influenced by ambient temperature, solar radiation and wind conditions prior to the spill.

In addition to energy-exchange mechanisms between the pool and its surroundings, the following parameters are also known to influence hazard zones:

- Atmospheric conditions (temperature, pressure, wind speed, humidity, solar radiation and cloud cover);
- Initial ground temperature;
- Physical properties of the material (vapor pressure, specific heat, latent heat of vaporization, density, viscosity);
- Spill surface geometry (roughness, gradient, obstructions);
- Percolation of the liquid into the soil;
- Viscous effects on spreading;
- Solubility in or reactivity of the liquid with water;
- Physical properties of the surface material (density, thermal conductivity, etc.); and
- Effect of ice formation in moist soil on the thermal diffusivity and conductivity of cryogenic liquid spills.

2.4.4. Flashing, Mixed Liquid–Vapor Releases

For liquids stored under pressure at a temperature above their atmospheric boiling points, a discharge to the atmosphere may result in a flashing, two-phase, mixed liquid–vapor release. If the flash to a liquid–vapor mixture occurs before the fluid has left the discharge pipe, flow rates can be lower, by a factor of 4 to 5, than rates for a similar nonflashing liquid release. This potential flashing, which is an important consideration in sizing emergency relief vents, has been addressed by others (Fletcher, 1983; Fletcher and Johnson, 1984; Fauske and Epstein, 1988).

Other factors that affect the discharge of a flashing, two-phase, liquid–vapor release include:

- Location of the hole in the liquid or vapor space;
- Foaming or nonfoaming fluids (foaming occurs with low surface tension liquids);
- Ratio of hole size to vessel diameter; and
- Liquid height-to-vessel diameter ratio.

If a vapor escapes from a leak in the vapor space of a vessel containing a pressurized liquid, the liquid remaining in the vessel is cooled or autorefrigerates. The escaping vapor is generated from the liquid mass by using the sensible heat contained in the liquid mass to supply the needed latent heat required for vaporization. The energy or heat lost to the escaping vapor by the liquid mass causes its temperature and the corresponding saturation pressure to drop. In the absence of external heating mechanisms, the pressure in the vessel drops to ambient pressure and the flow of vapor through the hole may be significantly reduced. However, the effect of the falling temperature may result in the embrittlement of the vessel's shell and its possible collapse, releasing the remaining cold liquid inventory which could then be vaporized by heat or energy from the surrounding pool.

2.4.5. Behavior of Flashing, Mixed Liquid–Vapor Releases

Immediately beyond the restricted discharge path in a liquid or two-phase liquid–vapor release, the liquid phase will break up into droplets. This droplet formation is governed both by shear forces at the liquid–vapor interface (mechanical break-up) and forces that result from the large volumetric expansion of a boiling liquid (flash atomization). Generally, where a greater fraction of liquid flashes to a vapor, smaller droplets will be formed. Investigation of droplet formation, size distributions, and their behavior in flashing liquids is described in Brown and York (1962), Wheatley (1986), and Lantzy et al. (1990).

Beyond the point of liquid droplet formation, ambient air will be entrained into the resulting two-phase jet, and the liquid droplets will begin to evaporate. Both the liquid droplets and the bulk liquid–vapor cloud, being heavier than air, will tend to settle toward the ground, but not necessarily at the same velocity. Large liquid droplets with sufficient downward settling velocity may rain out of the cloud onto the ground and form a liquid pool which then evaporates. Investigation of liquid rain-out in these types of releases has been conducted experimentally (Energy Analysts Inc., 1990; Lantzy et al., 1990). A number of numerical models (Iannello et al., 1988; Melhem and Saini, 1992; Woodward and Papadourakis, 1991), as well as empirical correlations (DeVaull and King, 1992), have been developed for predicting liquid rain-out in these releases. At this time, engineering models for predicting rain-out are still under development.

2.4.6. Gases/Vapors

The release rate of gases and vapors depends on the size and geometry of the hole, the flow path of the release stream, the storage pressure and temperature, and ambient pressure. When a gas or vapor is released across a flow restriction, it expands and its density changes. In making flow estimates for gases and vapors, this density change must be taken into consideration. Closed-form equations for estimating vapor flow rates may be found in many introductory fluid dynamics texts.

With a ratio between storage pressure and ambient pressure of about 2 or greater (for air), the flow rate is limited to the sonic velocity of the fluid at the end of the flow restriction (choked flow). At this point, the fluid pressure can be greater than the ambient pressure. The remaining expansion occurs beyond the flow restriction, where the release accelerates both radially and axially.

2.5. Consequences of a Release

An understanding of the potential consequences of a release is important in the design of mitigation strategies. For example, is it better to allow a pool to burn or to evaporate slowly? The consequences of a release depend on the possible hazard (fire, explosion, toxicity), the area affected by the hazard, and the level of damage or injury that would result. Approaches and techniques for consequence calculation are covered in detail in other CCPS publications, such as *Guidelines for Chemical Process Quantitative Risk Analysis* (CCPS, 1989a), *Guidelines for Use of Vapor Cloud Dispersion Models, Second Edition* (CCPS, 1996), and *Guidelines for Evaluating the Characteristics of Vapor Cloud Explosions, Flash Fires, and BLEVEs* (CCPS, 1994).

The severity and extent of any consequences depend on a number of factors:

- For a fire: thermal radiation, heat flux, and the duration of exposure determine the severity of injury or damage.
- For explosion overpressure: the magnitude of the impulse determines the severity of damage to structures and people. Generally the most severe injuries to people result from collapse of structures and flying debris.
- For a particular toxic material: the concentration and period of exposure (dosage) determine the hazard level. Data on hazards in some cases can be found in many documents, such as the Material Safety Data Sheets (MSDS) published by the manufacturer of the particular materials, in National Institute Occupational Safety and Health (NIOSH) publication NIOSH *Pocket Guide to Hazardous Chemicals* (DHHS, 1990), or *Emergency Response Planning Guidelines for Air Contaminants* (ERPGs) issued by the American Industrial Hygiene Association (1992).

2.5.1. Nature of Hazards

The postulation of a logical sequence of events leading to a potential hazard can be described using an event tree (CCPS, 1992), which represents potential pathways leading to many possible hazardous outcomes requiring quantification and mitigation. Event trees generally point to the following hazards: thermal radiation, exposure to toxic materials, explosion overpressures, and projectiles.

The event trees depicted by Figures 2.2 and 2.3 represent four typical loss of containment scenarios leading to hazards:

1. Finite duration liquid spill at or below its bubble point at atmospheric pressure;
2. Finite duration release of a subcooled, saturated, superheated or two-phase fluid at a temperature above its bubble point at atmospheric pressure;
3. Finite duration release of a gas/vapor from a vessel/source containing gas or a two-phase mixture; and
4. Catastrophic failure of a vessel containing a two-phase/saturated liquid or gas/vapor under pressure.

The logic of event trees is usually built into the structure of detailed consequence and risk models (CCPS, 1989a; Chemical Manufacturers Association, 1994). Typically the user specifies a list of scenarios where one or more of the release events described by 1 through 4 above are triggered.

For example, if we consider the release of refrigerated LNG from a large vessel, the following scenario outcomes are specified (see Figure 2.3):

1. The formation of a liquid pool which spreads and vaporizes as a function of time.
2. If ignition is immediate, the spill results in a pool fire and thermal hazard radiation footprints are established.
3. If ignition is delayed, the pool vaporizes and forms a flammable cloud. If ignition is encountered downwind, the vapor cloud can ignite and burn back to the source in the form of a vapor cloud fire and, if the release is still occurring, will cause a pool fire. If the flame encounters turbulence or significant blockage as it burns back to the source, a vapor cloud explosion can result.

The event trees represented in Figures 2.2 and 2.3 are most effectively utilized when they are automated so that the output of one or more models is fed directly into subsequent models. The availability of event trees in consequence analysis tools enables the user to efficiently perform sensitivity and what-if analysis as well as mitigation effectiveness assessment.

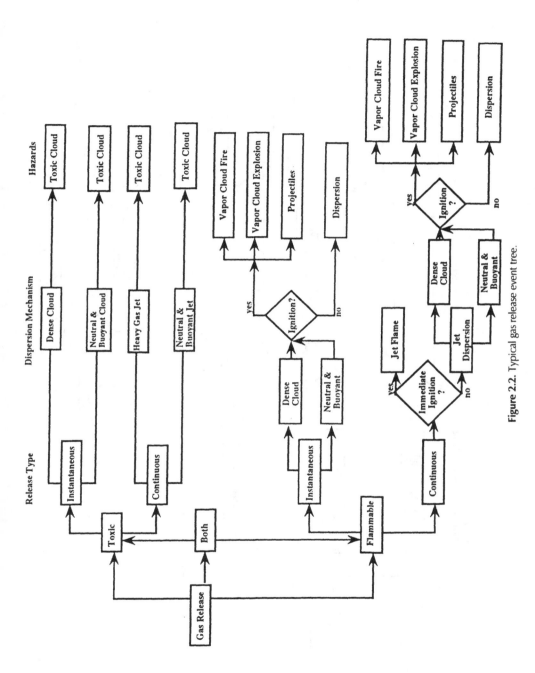

Figure 2.2. Typical gas release event tree.

19

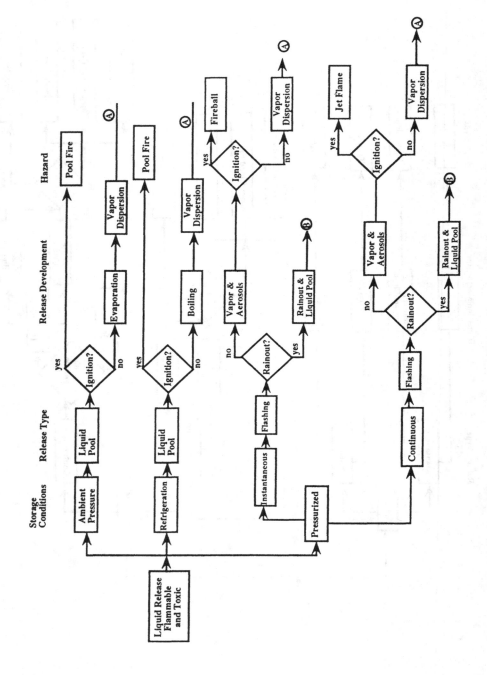

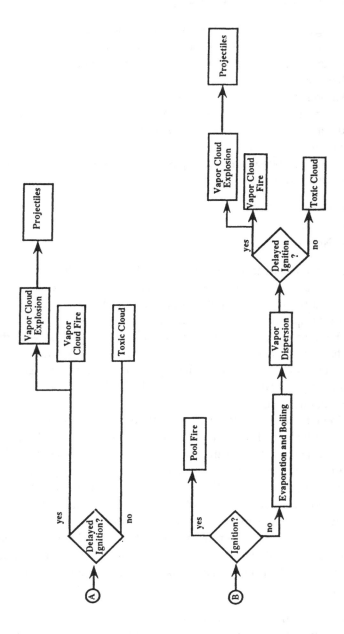

Figure 2.3. Typical liquid release event tree.

2.5.2. Toxic and Flammable Dispersion

Following their release, toxic and/or flammable materials that become airborne are carried by the wind and transported away from the spill site. While being transported downwind, the airborne chemical(s) mix with air and disperse.

Among the models required for hazard assessments, vapor dispersion models are perhaps the most complex. This is primarily because of the varied nature of both the release scenarios and the chemicals that may enter into the environment.

In vapor dispersion analysis, gases and two-phase liquid–vapor mixtures are divided into three general classes:

- Positively buoyant
- Neutrally buoyant
- Negatively buoyant.

These classifications are based on the density difference between the released material and its surrounding medium (air). The classifications are influenced by release temperature, molecular weight, presence of aerosols, ambient temperature at release, and relative humidity.

The density of the cloud or plume and its interaction with the ambient air affect the dispersion process. A buoyant cloud has a density less than that of the surrounding air. A heavy cloud has a density greater than the surrounding air. The density of a neutrally buoyant cloud is equal to that of the surrounding air.

Upon release, a buoyant material usually results in a rising plume, whereas a heavier-than-air material usually slumps toward the ground. In a release of heavier-than-air material at or near ground level, the initial density of the material determines its initial spreading rate. This is particularly true for large releases of liquefied or pressurized chemicals where flashing of vapor and formation of liquid aerosols contribute significantly to the effective vapor density and, therefore, to the density difference with air. Results of recent research programs (Melhem and Croce, 1994) demonstrate the importance of heavy-gas dispersion in the area of chemical hazard assessment. In fact, heavy-gas dispersion phenomena exhibit a predictable pattern:

- The initial rate of spreading (often termed slumping) of a heavier-than-air vapor cloud can be significant, depending on the magnitude of the difference between the effective mean cloud/plume density and the air density.

- The rapid mixing with ambient air caused by slumping enhances horizontal spreading beyond that seen in neutrally buoyant clouds.
- Because mixing in the vertical direction is suppressed by a stable density stratification, a slowly diluting vapor cloud that hugs the ground is generated.
- When the mean density difference becomes small, the subsequent dispersion is governed by prevailing atmospheric conditions.

Heavy-gas behavior is often dominant close to the point of release and in the near field. It is particularly important when considering large releases of pressurized or refrigerated flammable materials, for which the value of the lower flammable limit is low. Typical hydrocarbons that fall into this grouping are ethane, 2.9% by vol LFL; propane, 2.1% by vol LFL; and the butanes, 1.8% by vol LFL (Gas Processors Suppliers Association, 1972). Farther downwind, after additional mixing with air, the concentration of the flammable material is less important because it is then less than the lower flammable limit. The other hazard of heavy gases is asphyxiation of personnel who may inadvertently enter or be surrounded by the cloud.

Heavy-gas effects at the source are more important if the cloud's potential energy is larger than the turbulent kinetic energy generated by the surrounding air. Therefore, heavy-gas effects are more important at low wind speeds and stable atmospheric conditions than at high wind speeds and neutral or unstable atmospheric conditions.

Dispersion depends strongly on release modes. Depending on the nature of the release, the dispersion will occur in one or more of the following ways:

- An instantaneous or short duration release (puff);
- A continuous release (plume);
- A momentum-dominated continuous release (jet);
- A variable-rate, low-momentum release (plume); or
- A variable-rate momentum release (jet);

Each release mode introduces its own characteristics to the subsequent dispersion. For instance, a momentum-dominated jet will entrain air and dilute much faster than a low-momentum plume at the same release rate.

In addition to the influence of the initial release density, the presence of liquid in the cloud, release rate/quantity, release duration, and release mode, dispersion is also affected by:

- Prevailing atmospheric conditions,
- Elevation of the source,

- Surrounding terrain,
- Source geometry, and
- Direction of the release.

The lower the concentration of interest, the larger the dispersion distance. As with the source release rate, the effect is not linear.

The elevation of a source refers to its physical height above the reference point of interest, which in most cases is the ground. Unlike a ground-level release, an elevated release can be diluted from the top and bottom, which enhances dilution, particularly in the near field. From an elevated source, the peak ground-level concentration for a heavy or neutrally buoyant gas will be found some distance downwind from the source, in the area where the plume profile begins to touch the ground. In some cases, under stable atmospheric conditions if the plume is sufficiently elevated, significant ground-level concentrations may never occur.

Numerous dispersion models have been documented over the last two decades. These models can be divided into four general classes:

- Jet dispersion;
- Gaussian models for positively or neutrally buoyant clouds;
- Heavy-gas box, refined box, and slab models for heavy-gas dispersion; and
- Three-dimensional hydrodynamic models which can model complex releases and downwind obstructions.

All models possess empirical features and can be of questionable reliability under some release conditions. The Gaussian models are simple and valid for releases of neutrally or positively buoyant materials in a uniform flow field with no downwind obstacles. The box models represent a macroscopic approach to heavy-gas dispersion.

Three-dimensional models can handle complex cases with fewer empirical features. They require the solution of multidimensional, partial differential equations. These solutions are numerical tools and require the use of a supercomputer, or very long computation times on less sophisticated hardware.

2.5.3. Thermal Radiation

Thermal radiation hazards result from liquid hydrocarbon pool fires, flash fires, turbulent jet fires, and fireballs (BLEVE). A release may be ignited immediately or some time later, and the ignition source may be at the point of release or at a distance downwind, as shown in Figure 2.2. Gas venting with immediate ignition results in a jet flame if the release comes from a limited opening. Delayed ignition of a vapor release can result in a flash fire or explosion. If a liquid is released to the atmosphere and ignited immediately, a pool fire, which may be either confined (by a dike) or unconfined, will result.

If the material released to the atmosphere is not ignited, the spill can be accompanied by flash vaporization, liquid entrainment, and/or liquid accumulation (with pool formation and evaporation), and associated vapor dispersion. Absence of an immediate ignition source allows a vapor cloud to form as the vapors disperse downwind. A portion of this vapor cloud may be flammable, and if the gas has any toxic components, it can also pose a toxic hazard. The downwind extent of the flammable hazard depends on the size of the release, the upper and lower flammability limits of the material, and the air entrainment rate.

A remote ignition source could ignite the cloud, resulting in a fireball (mentioned above) or in a vapor-cloud fire that will burn from the point of ignition back toward the source of the cloud, that is, to the release point. The transient burning of the vapor cloud back to the release point could also initiate a subsequent, steadily burning pool fire, or a jet flame, depending on the source of the vapor cloud. If the amount of material in the flammable region of the cloud is large enough to enable flame acceleration, and if the ignition source is sufficiently strong, a vapor-cloud explosion may occur. Flame acceleration depends on the degree of confinement, turbulence and mixing, fuel reactivity, and the fuel–air ratio. Explosion hazards may also result from detonations of solid- and condensed-phase (not a vapor) highly reactive materials, as well as from the rupture of pressurized containers. Table 2.1 lists some ground-level values of thermal radiation that may cause injury or damage to materials such as steel, wood, and plastic (Theodore et al., 1989).

Products of combustion are sometimes hazardous. If this is the case for the material under consideration, the dispersion of the resulting toxic cloud must also be studied to determine the total consequences of the release.

If a pressure vessel is heated by an external fire or an internal runaway reaction, any liquid within the vessel will be heated, the pressure will increase, the vessel wall may weaken, or the vessel may become overpressurized because of an upset in normal operation. These conditions may result in a sudden catastrophic failure of the vessel during which the vessel energy will be released and a significant portion of the contained liquid vaporized. This results in an energetic, vigorously mixed, two-phase release that can produce an overpressure capable of causing structural damage to a surrounding building(s) or equipment. If the liquid is flammable and ignited on release, a fireball or an explosion can occur.

TABLE 2.1
Damage Caused at Various Incident Levels of Thermal Radiation
(Theodore et al., 1989)

Incident Flux (kW/m^2)	Type of Damage Caused[a]
37.5	Sufficient to cause damage to process equipment: 100% lethality
25.0	Minimum energy required to ignite wood at infinitely long exposures (nonpiloted): 100% lethality
12.5	Minimum energy required for piloted ignition of wood, melting plastic tubing: 100% lethality
4.0	Sufficient to cause pain to personnel if unable to reach cover within 20 seconds; blistering of skin (first degree burns are likely): 0% lethality
1.6	Will cause no discomfort over long exposure

[a]At the lower levels, where time is required to cause serious injury to people, there is often the possibility to escape or take shelter. The accuracy of the incident flux damage relationship is considered to be adequate for initial hazard assessments and within the estimation of hazardous incidents. The correlation of thermally induced damage or injury may be applied to hazard assessment.

There are two possible mechanisms that can lead to catastrophic vessel failure upon rapid depressurization, which is typically caused by metal failure or relief device actuation (Melhem et al., 1994):

- Spontaneous nucleation at the thermodynamic stability limit, i.e., the superheat limit
- Cavitation caused by homogeneous nucleation at temperatures lower than the superheat limit.

The conditions of temperature, pressure, tank metallurgy, pressure-relief device, and the level of liquid in the vessel interact to determine which of the two mechanisms will occur when a vessel is exposed to fire loading.

The superheat limit was first proposed in the late 1970s as a possible mechanism for explaining catastrophic vessel failures. Reid (1979), Jones (1985), Martinsen et al. (1986), Davenport (1988), and Dunn (1988) suggest that BLEVEs (Boiling Liquid Expanding Vapor Explosions) are superheat explosions and therefore are easily predicted by assessing the superheat limit for any pressurized liquid material.

Catastrophic vessel failures have also been reported at temperatures lower than the superheat limit. The experiments conducted by Melhem et al. (1993), Birk et al. (1993), and Ogiso et al. (1972) clearly show that vessel failures have occurred at small superheats. These failures can be attributed to very large rates of pressure drop leading to a metastable liquid and subsequent cavitation (bubble collapse) at temperatures below the superheat limit. Venart (1991) reports on a similar phenomenon which he describes as a boiling liquid, compressed bubble explosion (BLCBE).

The rate of pressure drop required to cause liquid metastablity will decrease as the critical temperature is approached (Elias and Chambre, 1993). A key variable to consider in pressure relief design is the rate of pressure decrease induced by rupture disk or relief device actuation. Ogiso et al. (1972) conducted small-scale experiments with superheated water to investigate the impact of large rates of pressure drop on repressurization. Table 2.2 illustrates the impact of having a large rupture disk (i.e., large rates of pressure drop) on vessel repressurization. An oversized rupture disk may not necessarily be safer. Safety depends on where the temperature that corresponds to the rupture disk set pressure is on the stability curve. This temperature should not be allowed to exceed the superheat limit (Reid, 1979). For reacting mixtures the superheat limit must be assessed for the mixture composition present in the vessel at the time of the relief device actuation. Relief systems should be designed to prevent rapid pressure decays.

2.5.4. Explosions

An explosion is a fast energy-release process. Typical sources of explosion energy include chemical and internal energy. Chemical energy is usually associated with combustion, polymerization, or decomposition reaction for gases, liquids, dusts, and solids. Internal energy is released from pressurized gases or liquids as pressure-volume or expansion energy when high-pressure

TABLE 2.2
Impact of Rupture Disk Size on Vessel Repressurization for Water at 130°C. Data taken from Ogiso et al. (1972)

Hole Diameter (mm)	Relative Diameter (Hole/Vessel)	Relative Area (Hole/Vessel)	Observation
50	1	1	
37.5	3/4	1/2	Cavitation and
25.0	1/2	1/4	overpressure
12.5	1/4	1/16	
6.7	1/7	1/56	Overpressure
3.2	1/16	1/244	No overpressure

storage vessels fail catastrophically. Explosions can occur in confined or partially confined conditions, such as in a building or in process vessels and pipings.

For explosions where rapid combustion is the energy-release process, a given mass of fuel-oxidant has a fixed amount of chemical energy. This amount of energy is released at different rates, depending on how fast combustion proceeds. Once released, chemical energy is converted to internal energy. As a result, temperature and pressure usually increase. For an adiabatic, closed, constant-volume system, the final equilibrium state is independent of the rate of energy release, and is a function only of the total amount of energy released. Depending on how fast combustion is proceeding—that is, flame speed—local overpressure is generated before reaching equilibrium conditions. The presence of obstructions or semiconfinement tends to increase turbulence or flame area, that leads to even higher flame speeds. A faster flame speed yields a faster local pressure rise and a higher local overpressure. Flame speed is a key factor in the hazard assessment of an explosion.

Depending on the flame speed, explosions can be classified as deflagrations or detonations. Deflagrations are characterized by the release of thermal energy caused by the flame moving at subsonic speeds in the unburnt medium. Energy is released by chemical reaction (combustion) initiated by heat transferred from the reaction zone to the unburnt material. Detonations are characterized by a flame front linked to a shock wave followed closely by a combustion wave. The flame front travels at sonic velocities in the burnt-gas medium and at supersonic velocities in the unburnt-gas medium. Energy is released by chemical reaction (combustion) initiated by compression (caused by the shock wave) of the unburnt materials to levels above their autoignition temperatures.

Deflagrations yield overpressures on the order of less than 8 to 10 times the initial pressure for most hydrocarbon–air mixtures. Higher overpressures have been observed in hydrocarbon–oxygen mixtures (Shepherd et al., 1991). Detonations yield overpressures on the order of 20 or more times the initial pressure. Typical detonation velocities are on the order of 2000 m/s.

A detonation can be produced by direct initiation or by a *deflagration-to-detonation transition* (DDT). The direct initiation of a detonation requires deposition of a large amount of energy into the explosive medium in a very short time (on the order of microseconds). Existing low-pressure data indicate that the transition of a deflagration to a detonation is possible if any of the following occur:

- The deflagration is proceeding at a sufficiently high flame speed—500 to 600 m/s (Shepherd et al., 1991).

- Turbulence is enhanced by the presence of obstructions (girders, pipes, etc.), strong turbulent jets, a long tube, or confinement.
- The detonation cell size (the minimum volume in which the heat release effect from chemical reaction exceeds the gas expansion effects) is smaller than the minimum diameter of the tube or distance between obstacles (Glassman, 1987).

2.5.5. Explosion Hazards

Hazards associated with explosions are classified as:

- Overpressure (blast effects);
- Projectiles (missiles);
- Thermal radiation from the heat generated by the burned gases;
- Generation of toxic materials; and
- Boiling liquid expanding vapor explosion (BLEVE) or boiling liquid compressed bubble explosion (BLCBE), discussed in Section 2.5.3 and in *Guidelines for Evaluating the Characteristics of Vapor Cloud Explosions, Flash Fires, and BLEVEs* (CCPS, 1994).

Structural damage caused by overpressure involves:

- Pressure–time histories caused by the explosion;
- Structural loading mechanisms;
- Yielding of structural members; and
- Dynamic response characteristics of the structure.

Pressure–time histories caused by explosions may be nonuniform and subject to amplification because of secondary shocks and shock reflections. Current models can provide only one- or two-dimensional histories. Failure modes are typically permanent deformation (plastic deformation/buckling), stable cracking (leaking), and brittle failure. Table 2.3 (Theodore et al., 1989) describes expected damage estimates for humans, structural elements, and process equipment for particular overpressures.

In addition to the composition, temperature, pressure, phase, and flammability limits, the following parameters influence the resulting overpressure:

- Time elapsed between release and ignition;
- Mass within the flammable limits;
- Presence of obstacles;
- Venting of gases generated by the explosion (for explosions confined in vessels or other structures);
- Ignition source strength;

TABLE 2.3

Explosion Overpressure Damage Estimates (Theodore et al., 1989)

Overpressure (psi)	Expected Damage
0.03	Occasional breaking of large windows already under stress
0.04	Loud noise (143 dB), sonic boom glass failure
0.1	Breakage of small windows under strain
0.15	Typical pressure for glass failure
0.3	Some damage to house ceilings; 10% window glass breakage
0.4	Limited minor structural damage
0.5–1	Windows usually shattered with some window frame damage
0.7	Minor damage to home structures
1	Partial demolition of houses, which are made uninhabitable
1–2	Corrugated metal panels fail and buckle
1–8	Range for slight to serious injuries caused by skin lacerations from flying glass and other missiles
1.3	Steel frame of clad building slightly distorted
2	Partial collapse of walls and roofs of houses
2–3	Nonreinforced concrete or cinder block walls shattered
2.3	Lower limit of serious structural damage
2.4–12.2	Range for 1–90% eardrum rupture among exposed population
2.5	50% destruction of brickwork
3	Steel frame distorted and pulled away from foundation
3–4	Frameless steel panel building ruined
4	Cladding of light industrial buildings ruptured
5	Wooden utility poles snapped
5–7	Nearly complete destruction of houses
7	Loaded train wagons overturned
7–8	8–12-inch-thick nonreinforced brick failure by shearing or flexure
9	Loaded train boxcars demolished
10	Probable total building destruction
15.5–29	Range for 1 to 99% fatalities among exposed population because of direct blast effects

- Laminar burning velocity and flame acceleration; and
- Type of fuel.

Additional details and references for explosion modeling maybe found in Baker et al. (1978), Baker et al. (1983), CCPS (1994); Lees (1980); Strehlow (1980), and Wiekema (1979).

2.6. Postrelease Mitigation Techniques

Those who respond to a release are usually heading into an area from which everyone else is trying to escape. Well-designed postrelease mitigation measures can provide response personnel with safer access to the hazard zone. Off-site consequences can be significantly reduced, as well.

Postrelease mitigation measures are intended to reduce the consequences of a release. They are designed to minimize the impact of an uncontrolled situation by containing the release in as small an area as possible and constraining the growth of the hazard zone. This subsection describes generic mechanisms of consequence reduction from which mitigation techniques are derived. All of these techniques are described in detail in subsequent chapters. Postrelease mitigation techniques fall into two categories: (1) containment or suppression, which involves limiting the amount of material that is released into the air, and (2) countermeasures, which are applied once the released material has formed a pool and/or a vapor cloud.

2.6.1. Containment or Suppression to Limit Releases to the Air

When a flammable or toxic material is released, all the potential hazards, except for pool and jet fires, are associated with airborne concentrations of the material. The material is either released as a vapor, subsequently vaporizes from a pool of spilled material, or is entrained as an aerosol during the release and subsequently vaporizes. This section considers methods for suppressing aerosol entrainment and evaporation.

The primary parameters affecting entrainment and evaporation are solar radiation, ambient temperature, storage or process temperature and pressure, liquid spill surface area, wind speed and the properties of the spilled material (such as vapor pressure, surface tension, and viscosity). Suppressing volatility by reducing containment temperature and constraining pool size or exposed surface area via a dike or berm are effective postmitigation approaches and are discussed in Chapters 3 and 5.

2.6.1.1. Secondary Release Containment
Secondary containment is a design technique that can be applied to vapor or liquid releases. Examples include a double-walled storage vessel (tank within a tank) and enclosures that surround a hazardous area, such as rail or truck loading/unloading racks. The design and operation of the latter must address the safety of operating personnel required to be inside the enclosure, as well as the containment of the release. Containment of liquid spills through properly designed curbing, dikes, and sumps is one of the most effective forms of postrelease mitigation. Another advantage of a liquid-containment system is that it is passive (except for the neutralization/disposal step) and requires minimal preventative maintenance.

2.6.1.2. Dilution
Dilution is used to mitigate hazardous releases, especially releases that form liquid pools. If properly implemented, the addition of an appropriate diluent to a liquid spill can have beneficial effects. Dilution can be employed to reduce the vapor pressure of the spilled hazardous material. It may also chemically combine with the hazardous material and render it nonhazardous. An example of this would be the use of water to dilute spills of a water-soluble material, such as acetone. Acetone is very soluble in water, and as it is diluted, the vapor pressure of the acetone above the mixture is reduced, commensurately reducing the potential for ignition and fire. However, to be effective there must be adequate volume available to obtain a satisfactory dilution and a way to dispose of the acetone–water mixture in an environmentally sound manner after the initial crisis has passed.

The effectiveness of dilution is highly chemical-specific and must be well conceived and tested. For instance, the improper addition of another material to a spill could cause the rapid evolution of heat, which could accelerate the evaporation rate and make the situation worse. Even with the proper diluent selection, the method and rate of addition are both important for controlling vapor evaporation rates. If water is to be used as a diluent for a material with a high heat of dilution, such as sulfuric acid or oleum, it must be added rapidly and in a large enough quantity to not only effect the dilution, but also serve as a heat sink for the heat of dilution produced. In this way, high evaporation rates for the pool being diluted will be minimized.

In certain cases, adding material (e.g., a weak base) to neutralize the spilled material (acidic) is appropriate. This approach relies on the chemical destruction of the hazardous material and requires a good understanding of the material's chemistry.

2.6.1.3. Reducing the Evaporation Surface Area

After a liquid spill it may be possible to reduce the surface area by covering the spill or designing drainage systems to minimize surface area. These techniques work by reducing the evaporative surface area.

Fire-fighting foams which do not break down or dry powders can be effectively applied to spills within dikes or on the open ground. Sheeting can be applied over the surface of a dike to keep the surface cool. Layers of floating balls can be applied within dikes, but unless at least three layers can be established (Feind, 1975), there is a danger that vaporization will be enhanced. This occurs by providing a larger evaporative surface as turbulence causes the balls to become fully wetted. Special considerations are required to ensure the integrity of these systems during high wind conditions to assure the balls don't blow to one side or the foam is blown off the surface it is covering. These techniques are not generally applicable to cold liquids, as the warm cover material will initially increase vaporization.

Sloped dike floors or sumps within the dike can also be used to reduce the surface area of the spill and allow the materials to be transferred to emergency holding systems.

2.6.1.4. Momentum Breaks

If aerosolization is a major contributor to a hazard, a portion of the airborne liquid can be removed by placing an angled plate over the release point. The released material strikes the plate, which disperses much of the momentum and enhances coagulation of liquid droplets. In this way a portion of the liquid will drop out of the cloud. However, the loss of momentum may result in less efficient vapor dispersion and, hence, a larger hazard area.

This approach is practical only when the release point is predictable, for example, at a vent pipe or pump seal, as the plate must be installed before a release occurs. Furthermore, the release must be carefully modeled to ensure that a smaller hazard area will result.

Another more common example of a momentum break is the use of flange guards placed around flange joints in piping systems handling sulfuric acid or strong caustic. They knock down the liquid spray from a small leak, to prevent personnel injury, and also change color to indicate the leak location.

2.6.2. Countermeasures

The type of countermeasure that will be effective depends on whether the initial release is all vapor, a vapor–liquid mixture (aerosol), or forms a pool.

2.6.2.1. Chemical Absorption of Vapor Cloud

Some toxic chemicals have a strong affinity for water or other liquids. Sprays in the path of the cloud will absorb some of the chemical and will induce more turbulence in the cloud which can lead to enhanced dispersion. Water is used for vapors that have an affinity for water, for example halogen acids, and ammonia.

2.6.2.2. Plume Dilution

Attempts can be made to minimize the impact area of hazardous vapor releases. This can be accomplished by taking steps to enhance near-field vapor dispersion and thus increase the dilution rate to break down aerosols, or to move the cloud away from sensitive areas. The techniques listed below have all been used:

- Air and steam curtains, which can also act as a physical barrier to the vapor cloud and prevent its migration towards sensitive locations;
- Water sprays and curtains, which also provide physical barriers and may absorb hydrophilic chemicals;
- Fans and eductors;
- Thermal curtains (fire trenches), which may destroy chemicals that are flammable or thermally unstable and provide an upward draft or air current;
- Vapor barriers or vapor fences to contain dense, evaporated vapors and promote dispersion; and
- When possible, point vents from relief valves and bursting disc so they will discharge in a direction away from critical areas.

2.6.2.3. Deliberate Ignition

A technique that is only occasionally used to control flammable vapors is deliberate ignition. This countermeasure is most beneficial when the released material is both flammable and highly toxic (e.g., hydrogen cyanide or acrolein) and in a location remote from people and property. Since the toxicity hazard zone is much larger than the flammability hazard, early ignition will eliminate much of the toxicity hazard. A drawback of this approach is that igniting the cloud may alleviate a downwind hazard but create a greater local hazard, such as an explosion. If this strategy is to be employed, it requires careful planning and consideration, beforehand.

2.7. References

American Industrial Hygiene Association. 1992. *Emergency Response Planning Guidelines for Air Contaminants.* Fairfax, VA: American Industrial Hygiene Association.

Baker, W. E., J. J. Kulesz, R. E. Richer, P. S. Westine, V. B. Parr, L. M. Vargas, and P. K. Moseley. 1978. *Workbook for Estimating the Effects of Accidental Explosions in Propellant Handling Systems.* NASA Report CR-3023. Lewis Research Centre. Washington, D.C.: National Aeronautics and Space Administration.

Baker, W. E., P. A. Coz, P. S. Westine, J. J. Kulsez, and R. A. Strehlow. 1983. *Explosion Hazards and Evaluation.* New York: Elsevier.

Birk, A. M., M. Cunningham, Z. Ye, and J. Maillette. 1993. Medium-Scale Fire Tests Investigating a BLEVE Event. In *Tenth Technical Seminar on Chemical Spills,* pp. 203–220. Environment Canada.

Brown, R. and J. L. York. 1962. *AIChE Journal,* 8(2): 149.

CCPS (Center for Chemical Process Safety). 1988a. *Guidelines for Vapor Release Mitigation.* New York: American Institute of Chemical Engineers.

CCPS (Center for Chemical Process Safety). 1988b. *Guidelines for Safe Storage and Handling of High Toxic Hazard Materials.* New York: American Institute of Chemical Engineers.

CCPS (Center for Chemical Process Safety). 1989a. *Guidelines for Chemical Process Quantitative Risk Analysis.* New York: American Institute of Chemical Engineers.

CCPS (Center for Chemical Process Safety). 1989b. *Guidelines for Technical Management of Chemical Process Safety.* New York: American Institute of Chemical Engineers.

CCPS (Center for Chemical Process Safety). 1992. *Guidelines for Hazard Evaluation Procedures, Second Edition with Worked Examples.* New York: American Institute of Chemical Engineers.

CCPS (Center for Chemical Process Safety). 1994. *Guidelines for Evaluating the Characteristics of Vapor Cloud Explosions, Flash Fires, and BLEVEs.* New York: American Institute of Chemical Engineers.

CCPS (Center for Chemical Process Safety). 1996. *Guidelines for Use of Vapor Cloud Dispersion Models, Second Edition.* New York: American Institute of Chemical Engineers.

Chemical Manufacturers Association (CMA). 1994. *Inter Industry Rail Safety Task Force: Transportation Risk Model.* Washington, D.C.: Chemical Manufacturers Association.

Davenport, J. A. 1988. Hazards and Protection of Pressure Storage and Transportation of LP Gas. *Journal of Hazardous Materials,* 20: 3–19.

DeVaull, G. E. and J. A. King. 1992. Similarity Scaling of Droplet Evaporation and Liquid Rain-out Following the Release of a Superheated Flashing Liquid to the Environment. In *Proceedings of the 85th Annual Meeting Air and Waste Management Association, Kansas City.* Pittsburgh: Air and Water Waste Management Association.

DHHS (NIOSH) (Department of Health and Human Services, National Institute Occupational Safety and Health). 1990. *NIOSH Pocket Guide to Chemical Hazards.* U. S. Department of Health and Human Services, Center for Disease Control, Cincinnati, Ohio.

Dunn, V. 1988. BLEVE: The Propane Cylinder. *Fire Engineering* 131:63–70.

Elias, E., and P. L. Chambre. 1993. Flashing Inception in Water during Rapid Decompression. *Journal of Heat Transfer* 115: 231–238.

Energy Analysts, Inc. 1990. Release Characteristics of Superheated Water and Freon-11 Liquids. Center for Chemical Process Safety, Experimental 90-05-540. New York: American Institute of Chemical Engineers.

Fauske, H. K. and M. Epstein. 1988. Source Term Considerations in Connection with Chemical Accidents and Vapor Cloud Modeling. *Journal of Loss Prevention, Process Industries*, 1: 75–83.

Feind, K. 1975. Reading Vapor Loss in Ammonia Tank Spills. *Safety in Ammonia Plants and Related Facilities Symposium, 17th Annual Proceedings*, pp. 114–118. American Institute of Chemical Engineers National Meeting, Salt Lake City.

Fletcher, B. 1983. Flashing through Orifices and Pipes. *AIChE Loss Prevention Symposium, Denver*. New York: American Institute of Chemical Engineers.

Fletcher, B., and A. E. Johnson. 1984. The Discharge of Superheated Liquids from Pipes. In *IChemE Symposium Series 85*. London: IChemE.

Fryer, L. S., and G. D. Kaiser, 1979. DENZ—A Computer Program for the Calculation of Dispersion and Dense Toxic of Explosive Gases in the Atmosphere. *UKAEA Safety and Reliability Directore Report*. SRD R152, Calcheth, UK.

Gas Processors Suppliers Association (GPSA). 1972. *Engineering Data Book*. 9th Edition. Tulsa, Oklahoma: Gas Processors Suppliers Association.

Glassman, I. 1987. *Combustion* 2nd Edition. Academic Press Inc. Orlando, Florida.

Iannello, V., G. B. Wallis, and P. H. Ruther. 1988. *Liquid Release Final Report*. Prepared by Creare, Inc. for the Center for Chemical Process Safety. New York: American Institute of Chemical Engineers.

Jones, M. R. 1985. Vapour Explosions Resulting from Rapid Depressurization of Liquids: The Importance of Initial Temperature. In *IChemE Symposium Series 93*, pp. 357–362. London: IChemE.

Lantzy, R. J., R. D. Myers, D. B. Pfenning, and S. B. Millsap. 1990. Atmospheric Release Tests of Monomethylamine. *Journal of Loss Prevention*, 3: 77.

Lees, F. P. 1980. *Loss Prevention in the Process Industries*. London and Boston: Butterworths.

Martinsen, W. E., D. W Johnson, and W. F. Terrel. 1986. BLEVEs: Their Causes Effects and Prevention. *Hydrocarbon Processing* 5(11): 141–148.

Melhem, G. A. and P. A. Croce. 1994. "Advanced Consequence Analysis: Emission, Dispersion, Fire and Explosion Dynamics." Working Manuscript, Arthur D. Little, Inc.

Melhem, G. A. and R. Saini. 1992. A Model for the Dispersion of Two-Phase Flashing Multicomponent Jets. *Proceedings of the Process Plant Safety Symposium, South Texas Section, Houston, Texas*. New York: American Institute of Chemical Engineers.

Melhem, G. A., P. A. Croce, and H. Abraham. 1993. Data Summary of the National Fire Protection Association's BLEVE Tests. *Process Safety Progress*, 12(2): 76–82.

Melhem, G. A., H. G. Fisher, and D. A. Shaw. 1994. An Advanced Method for the Estimation of Reaction Rates, Scale-up and Pressure Relief Design. *Proceedings of the 28th Loss Prevention Symposium*.

Ogiso, C., N. Takagi, and T. Kitagawa. 1972. On the Mechanism of Vapor Explosion. In *Loss Prevention and Safety, Session 9*, pp. 233–240, PACHEC.

Reid, R. C. 1979. Possible Mechanism for Pressurized-Liquid Tank Explosion or BLEVEs. *Science*, 203: 1263–1265.

Shepherd, J., G. A. Melhem, and P. Athens. 1991. Unconfined Vapor Cloud Explosions: A New Perspective. *International Conference and Workshop on Modeling and Mitigat-*

ing the Consequences of Accidental Releases of Hazardous Materials. New York: American Institute of Chemical Engineers.

Strehlow, R. A. 1980. Blast Wave from Deflagration Explosion in an Acoustic Approach, Loss Prevention. *Chemical Engineering Progress Technical Manual.* Vol. 14, pp. 145–153. New York: American Institute of Chemical Engineers.

Theodore, L., J. P. Reynolds, and F. B. Taylor. 1989. *Accident and Emergency Management.* New York: John Wiley & Sons.

Venart, J. E. S. 1991. To BLEVE or Not to BLEVE: Anatomy of a Boiling Liquid Expanding Vapor Explosion. New Brunswick: Fire Science Centre, University of New Brunswick. *Process Safety Progress,* April: 67–70

Wheatley, C. J. 1986. Factors Affecting Cloud Formation from Releases of Liquified Gases. *IChemE Symposium on Refinement of Estimates of the Consequences of Heavy Toxic Vapor Releases,* Manchester, UK. London: IChemE.

Wiekema, B. J. 1979. Vapor Cloud Explosions. In *Methods for the Calculation of the Physical Effects of the Escape of Dangerous Materials: Liquids and Gases* ("The Yellow Book"). P. O. Box 3432, 7300 AH Apeldoorn, The Netherlands.

Woodward, J. L. and A. Papadourakis. 1991. Modeling of Droplet Entrainment and Evaporation in a Dispersing Jet. *Proceedings of the International Conference and Workshop on Modeling and Mitigating the Consequences of Accidental Releases of Hazardous Materials.* New York: American Institute of Chemical Engineers.

3

Vaporization Reduction

3.1. Introduction

3.1.1 Why Reduce Vaporization Rates?

The release of a toxic or hazardous material usually results in minimum impact or presents little danger while it remains in a nonairborne liquid state in which it can be more easily contained or controlled. When any or all of the material vaporizes and becomes airborne, however, it becomes hazardous to people both close to the release and in surrounding areas. Therefore any steps that can be taken to reduce the quantity of toxic/flammable material that vaporizes will serve to reduce the consequences of the release and the total number of people involved.

The effects of the release of a toxic material are proportional to both its airborne concentration and the duration of the release; that is, the greater the concentration or time of exposure, the greater the consequences to those who are exposed. The toxic effect can be expressed mathematically in a probit equation that calculates the probability of damage, Y (CCPS, 1989a). The equation is

$$Y = A + B \ln[C^n t] \tag{3-1}$$

where A and B and n are constants.

Dose is the product of concentration, C, and time, t. As the equation shows, when n is greater than 1.0, as it is for many compounds, concentration effects are actually the more significant component of the total dose. Therefore, reducing the vaporization rate reduces the toxic dose because of its impact on reduced concentration, even though the lower vaporization rate may increase the duration of the exposure.

Values for the constants in the probit equation (3-1) for a few compounds can be found in *Guidelines for Chemical Process Quantitative Risk Analysis* (CCPS, 1989a, Table 2.9, p. 156).

Other means of evaluating the toxic effects of vapors, and their potential impact on exposed people, can be found in the *Emergency Response Planning Guidelines for Air Contaminants* (ERPGs), issued by the American Industrial Hygiene Association (1992). The documents define values of exposure to toxic vapors needed to produce specific health effects. Three levels of vapor concentration are defined:

- **ERPG-1** is the maximum airborne concentration below which it is believed that nearly all individuals could be exposed for up to 1 hour without experiencing other than mild, transient adverse health effects or perceiving a clearly defined, objectionable odor.
- **ERPG-2** is the maximum airborne concentration below which it is believed that nearly all individuals could be exposed for up to 1 hour without experiencing or developing irreversible or other serious health effects or symptoms that could impair an individual's ability to take protective action.
- **ERPG-3** is the maximum airborne concentration below which it is believed that nearly all individuals could be exposed for up to 1 hour without experiencing or developing life-threatening health effects.

When using ERPG data, keep in mind that human responses vary widely and that particular responses cannot be attributed to exact levels of exposure. Other sources of hazards data in some case can be found in Material Safety Data Sheets (MSDS) issued by the manufacturer of the material or in the National Institute of Occupation Safety and Health (NIOSH) publication NIOSH *Pocket Guide to Chemical Hazards* (DHHS, 1990).

In the case of flammable materials, reducing the vaporization rate of the material released tends to keep it below explosive or flammable limits, or at least minimizes the size of the flammable vapor cloud formed and, therefore, the potential damage from its ignition.

In this chapter we will be discussing methods to reduce the rate of vaporization of a released material to minimize its impact; however, all of the principles of good practice that have been presented in *Guidelines for Safe Storage and Handling of High Toxic Materials* (CCPS, 1988b) and *Guidelines for Technical Management of Chemical Process Safety* (CCPS, 1989b) still apply. The discussions in this guideline relate to cases in which unexpected, accidental failures could occur.

3.1.2. Methodology

3.1.2.1. Principles of Vaporization
Vaporization can be defined as the change of a substance from a liquid phase to a gas or vapor phase. The different mechanisms for converting a liquid into a vapor and the resulting consequences were described in Sections 2.4 and 2.5 of this guideline.

The major factors that affect vaporization include:

- Vapor pressure, or the boiling point, of the material, which will determine the following:
  —Whether the material will be a gas when released
  —Whether flashing will occur and thus provide energy for flash atomization
  —Whether all of the material will be a liquid which will then vaporize.

Which of these phenomena occurs depends on the substance and the process conditions before release, as well as temperature and pressure. After release, the following factors will affect vaporization:

- Surface area
  —The larger the surface area of the spilled material that accumulates as a liquid pool, the more material can evaporate or vaporize per unit of time.
- Wind velocity
  —The mass transfer coefficient for convective heat loss is proportional to the wind speed.
- Ground temperature
  —Due to heat conduction from the ground to the pool, a higher ground temperature will cause a higher initial vaporization rate.
- Available energy for vaporization
  —Part of the energy for vaporization comes from the substance when it is released; the balance of the needed energy comes from the surroundings. How fast this energy will be transferred depends on the temperature of the material, the temperature of the surroundings, and the amount of surface area available. The energy can come from either the ground on which the liquid pool lies, or from the atmosphere in the case of an aerosol formed during the event.

Consider the spill of a liquid mixture containing C components on a flat, horizontal spill surface. The resulting liquid pool is circular with a radius r and a liquid depth z_l (see Figure 3.1). Mass can leave the pool by

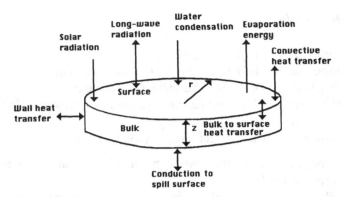

Figure 3.1. Pool heat and mass transfer mechanisms. (From Melhem and Croce, 1994.)

evaporation and percolation/dissolution into the spill surface. A mass balance can be written for each component (i) in the liquid pool (see Melhem and Croce, 1994):

$$\frac{dn_i}{dt} = \frac{dn_{i\,in}}{dt} - \frac{dN_{i\,out}}{dt} - \frac{dn_{i\,out}}{dt}$$ (3.2)

Accumulation = Input – Evaporation – Percolation/Dissolution

where N represents the number of moles (kmol) of vapor and n represents the number of moles (kmol) of liquid. The emission rate from liquid pools can either be established experimentally or estimated using detailed transient mass and energy balances. Typical experimental results are reported as a liquid regression rate, w_i. The number of moles lost due to evaporation is given by

$$\frac{dN_i}{dt} = \frac{\pi r^2 w_i \rho_{li}}{M_i}$$ (3.3)

where w_i is the liquid regression rate in (m/s), ρ_{li} is the liquid density of component i in (kg/m^3), r is the pool radius, and M_i is the molecular weight. The liquid regression rate is an important parameter in determining pool spreading behavior.

Detailed material and energy balance represent an alternate method for estimation of liquid pool emission rates. The mass transfer from the liquid pool surface to the surroundings gas phase can be estimated from the following equation:

$$\frac{dN_i}{dt} = k_{g,i} \pi r^2 (Y_i^* - Y_{o,i})$$ (3.4)

where $k_{g,i}$ is an overall gas phase mass transfer coefficient in (kmol/m²/s), r is the liquid pool radius in (m), Y_i^* is the vapor mole fraction in equilibrium with the liquid at the pool surface temperature, and Y_{oi}^* is the vapor mole fraction present in the surroundings at ambient conditions.

Determination of the value of $K_{g,i}$ is based on the experimental work of Mackay and Matsugu (1973):

$$k_{g,i} = 4.82 \times 10^{-3} \, u_w^{\left(\frac{2-n}{2+n}\right)} N_{Sc_i}^{-0.67} d^{\left(\frac{-n}{2+n}\right)} \left(\frac{P_s}{R_g T}\right) \qquad (3.5)$$

Here, $k_{g,i}$ is in (kmol/m²/s), P_s is the ambient pressure in (Pa), T is the liquid pool surface temperature (K), u_w is a 10 meter wind speed in (m/s), and n is an exponent that depends on atmospheric stability and surface roughness. Typical values for n range between 0.25 and 1. When n is equal to 0.25, the wind speed profile implied by the correlation is a 1/7 power law profile. For $n = 0.25$, equation (3-5) reduces to

$$k_{g,i} = 4.82 \times 10^{-3} u_w^{0.78} N_{Sc_i}^{-0.67} d^{-0.11} \left(\frac{P_s}{R_g T}\right) \qquad (3.6)$$

The Schmidt number N_{Sc_i} is defined as the ratio of momentum to mass diffusivity:

$$N_{Sc_i} = \frac{v_a}{D_{a,i}} \qquad (3.7)$$

where v_a is the kinematic viscosity of air in (m²/s) and $D_{a,i}$ is the diffusivity of component i in air, in (m²/s).

The overall mass phase coefficient $k_{g,i}$ also depends on the gas and liquid phase transfer coefficients:

$$k_{g,i} = \frac{1}{1/k_{v,i} + \Phi_{l,i}/k_{l,i}} \qquad (3.8)$$

where $k_{l,i}$ is the liquid phase mass transfer coefficient and $k_{v,i}$ is the vapor phase mass transfer coefficient, and $\Phi_{l,i}$ is the fugacity coefficient to account for liquid phase nonideality. When the liquid pool is a single component $k_{g,i} = k_{v,i}$ and $k_{g,i}$ may be determined by equation (3.6) above.

For situations in which the evaporation rate is limited by the liquid pool, consider the following:

$$y_i^* = \frac{\Phi_{l,i} x_i}{\Phi_{v,i}} \qquad (3.9)$$

where $y_i{}^*$ and x_i are the vapor and liquid phase mole fractions and $\Phi_{l,i}$ and $\Phi_{v,i}$ are the liquid and vapor phase fugacity coefficients.

For ideal liquid and vapor phase behavior

$$y_i^* = \frac{P_{\text{sat},i}\, x_i}{P_s} \qquad (3.10)$$

where $P_{\text{sat},i}$ is the saturation vapor pressure of the component and P_s is the total pressure. If $Y_{o,i}$ in equation (3.4) above equals 0, then equation (3.4) can be written as:

$$\frac{dN_i}{dt} = \pi r^2\, k_{g,i}\, x_i\, \frac{\Phi_{l,i}}{\Phi_{l,i}} \qquad (3.11)$$

For liquid pools containing a mixture of liquids of varying volatility, the light components will evaporate first. When evaporation rates are high, the surface concentration of the volatile components will be depleted. At this time the transfer by diffusion of these components from the bulk of the liquid to the surface becomes the limiting factor. The one-dimensional representation of vertical diffusion for a component i through a liquid layer of depth z_1 is given by:

$$\frac{\partial n_i}{\partial t} = -\left[\frac{D_i}{z_1}\right]\frac{\partial n_i}{\partial z} \qquad (3.12)$$

In this equation D_i is the molecular diffusion coefficient expressed as m^2/sec, and n_i is the number of liquid moles of component i. This equation can be simplified using approximate analytical solutions for the transient diffusion equation in the vertical direction to:

$$\frac{dn_i}{dt} = \frac{-0.632 n_i}{z_1}\sqrt{\frac{D_i}{t}} \qquad (3.13)$$

The liquid diffusion coefficient of component i in the mixture (D_i) can be estimated using a Wilke-Change method variation:

$$D_i = 1.1728 \times 10^{-15}\left(\frac{\sqrt{\sum_{j=1,\, j\neq i}^{C} x_j\, \phi_j M_{wj}}}{\mu_m v_i^{0.6}}\right) T_1 \qquad (3.14)$$

ϕ is an association factor which has a value of 1 for unassociated liquids and a value of 2.6 for water, 1.9 for methanol, and 1.5 for ethanol. T_1 is the liquid bulk temperature (K) which is different from the liquid surface temperature T. In these situations the rate-limiting process may change as a function of time from diffusion limited to mass transfer limited.

We divide the liquid pool into two regions: a bulk region and a surface region. Heat is transferred by conduction from the spill surface and from the dike or tank walls (if applicable) to the bulk liquid and then from the bulk liquid to the pool surface (see Melhem and Croce, 1994).

The surface temperature can be calculated using a comprehensive energy balance which accounts for heat loss due to evaporation, heat exchange due to surface conduction, solar heating and convection. The total pool energy exchange rate is:

$$\frac{dQ_s}{dt} = \frac{dQ_{sol}}{dt} + \frac{dQ_{con}}{dt} + \frac{dQ_{bulk}}{dt} - \frac{dQ_{evap}}{dt} + \frac{dQ_{cond}}{dt} \qquad (3.15)$$

where dQ_{sol}/dt is the heat rate gained by the pool from its exposure to solar flux, dQ_{bulk}/dt is the heat exchange rate between the bulk liquid and the pool surface, dQ_{evap}/dt is the energy loss rate from the pool due to evaporation/boiloff, dQ_{con}/dt is the convective heat transfer rate between the air and the pool, and dQ_{cond}/dt is the heat rate transferred to the pool surface from the air by water vapor condensation.

For the bulk liquid, the energy balance accounts for heat exchanged via ground conduction, conduction from tank or dike walls, etc. The unsteady energy balance of the bulk liquid can be completely described based on first principles by writing the equations describing the internal energy change. The overall internal energy change of the liquid pool is:

$$a\frac{dT_1}{dt} + b\frac{dP_s}{dt} = c \qquad (3.16)$$

where a, b, and c are constants defined as follows:

$$a = \sum_{j}^{C} n_i C_{vi} + n_T \frac{\partial \Delta U_v}{\partial T}$$

$$b = n_T \frac{\partial \Delta U_v}{\partial P}$$

$$c = \frac{dQ}{dt} - \sum_{i}^{C} \left(\int_{T_{ref}}^{T} C_{vi}\, dT + \Delta \underline{U}_v + n_T \frac{\partial \Delta U_v}{\partial n_i} \right) \frac{dn_i}{dt}$$

$$\frac{dQ}{dt} = \frac{dQ_{grd}}{dt} + \frac{dQ_{wall}}{dt} - \frac{dQ_{bulk}}{dt} - \underline{H}_1 \sum_i^c \frac{dn_{i,out}}{dt}$$

where $dQ_{grd/wall}/dt$ is the heat exchange rate due to conduction from the spill surface and dike/tank walls, C_v is the constant volume heat capacity (J/kmol/K), $\underline{U}$ is the molar internal energy, and $\underline{H}$ is the molar enthalpy.

For boiling pools, surplus energy is converted into boiling energy,

$$\sum_i^c \frac{dN_{i,out,b}}{dt} = \frac{Q_t}{\underline{\lambda}} \tag{3.17}$$

where $N_{i,out,b}$ is molar loss rate attributed to boiling (kmol/s) and $\underline{\lambda}$ is the molar latent heat of vaporization of the mixture.

Techniques that are successful in reducing vaporization will interfere in some way with one or more of the factors that affect the material and energy balance, or the rate of vaporization. In many cases, the final mitigation plan will not involve just one technique, but a combination of several. Some of the applicable techniques for reducing vaporization are described in the following subsections.

3.1.2.2. Reduction Measures

Thermal Controls. Cooling/refrigeration of a stored material reduces the energy available for converting the material to the vapor phase (discussed earlier), and it will reduce the vapor pressure. This is discussed in more detail in the following subsections. Some thermal controls include:

- Refrigerated storage;
- Shading the potential spill area, which limits solar radiation and, hence, reduces surface temperature and volatilization;
- Insulation of the spill surface; and
- Maintaining a small quantity of liquid (e.g., water) in a dike to cover the spill surface, which may significantly reduce heat conduction impacts during a spill event.

Containment. Keeping a released material confined is the next essential step in postrelease mitigation. This action limits the surface area and allows other mitigation techniques to be used in a more effective and efficient manner.

Covers. Putting a cover over a liquid pool is another technique for controlling a toxic or flammable material. Depending on conditions, the cover slows or prevents additional material from getting into the atmosphere.There are many types of covers; for example, floating roofs, foam, ping-pong balls, water or some other suitable fluid, or sheets of plastic.

Absorption.   If a material is absorbed when combined with another substance, its rate of vaporization will be greatly reduced or eliminated. This action can be accomplished by diluting the material with water, or introducing an absorbent or other material(s) onto which it binds. This measure is most effective for small spills.

Reactions.   When a hazardous material reacts with another substance, either it will become neutralized, or it will form a compound that is less volatile and ideally harmless. For example, reacting a spilled volatile acid with limestone placed around a storage tank or in a collecting sump will neutralize the acid and form a calcium salt. The neutralization reaction will generate heat that could just increase temperature or cause a short increase in vapor evolution.

Combustion is another reaction that can be used to control the release of a toxic gas. It is an attractive option, if the products of combustion are harmless, or the combustion results in a more expansive dispersion of the material. See Section 3.4.

3.1.2.3. Combinations

In actual practice, the measures described above, by themselves, may not reduce the rate of vaporization to the maximum extent possible. In most cases, combinations of these measures are used. One example is the release of a refrigerated material which keeps the hazardous material in liquid form. In such cases, the material is released to a diked area where it is confined. A cover of foam is then used to further reduce the rate of vaporization.

For each situation that has to be addressed, there are combinations of the techniques described above that will result in optimum vapor reduction. More details on these techniques are presented in the following sections.

3.2. Refrigeration

3.2.1. Effect of Refrigeration on Vaporization Rates

Refrigeration is a prerelease technique that can contribute to reducing the consequences of a release. It is useful when there are process reasons that require the use of a hazardous or toxic material at reduced temperatures, ideally, below its atmospheric boiling point. In such cases the refrigeration requirement can reduce the potential consequences of a release by reducing the system pressure, which affects the rate of flow, and the quantity of the material that remains as a liquid. The potential for refrigeration to reduce consequences and enhance the use of other postrelease mitigation techniques should be studied and the results reflected in the final design of the system. As shown in *Guidelines for Vapor Release Mitigation* (CCPS, 1988a) refrigeration of a material will yield the following benefits:

- The system pressure may be reduced so that the total flow through the leak is reduced.
- The potential effect of flashing to the vapor phase is reduced.
- The initial rate of vaporization from the pool may be reduced.
- The amount of material that is caught up in a mist is reduced by lowered jet entrainment because of the lower discharge pressure.

In addition to using refrigeration to maintain a liquid at lower temperatures and thus to mitigate the consequences of a release, there are situations where, once a liquid is released, the application of a cooling material to reduce the vaporization rate is a possibility (Greer and Gross, 1980; Greer, 1976). Studies of this technique have been conducted. Materials and substances that have been used to provide the cooling needed in this situation include ice or iced water, carbon dioxide as both a liquid and a solid, liquid air, and liquid nitrogen.

The major issue raised in these studies (Greer and Gross, 1980; Greer, 1976) was having the cooling material available in a timely manner. The cryogenic agent that seems most usable based on the subject studies is liquid carbon dioxide, which can be stored under pressure. In this case, it would probably serve a dual function for both postrelease mitigation and firefighting in the area covered.

Some tests conducted using the application of a cooling medium showed evaporation rate reductions on the order of 50%, just from the resulting reduction of the pool's temperature and the vapor pressure of the liquid. Naturally this type of reduction is totally dependent on the physical properties of the spilled material (Greer and Gross, 1980; Greer, 1976).

3.2.2. System Issues

Although the refrigeration of a hazardous process fluid will provide the benefits listed above, this technique does necessitate the insulation of the process equipment and piping. This makes it more difficult to inspect the equipment for external corrosion and embrittlement problems, unless there is an aggressive inspection program in place or noncorroding materials of construction are employed.

In addition, the reliability of the refrigeration equipment and the provision of spare machines and backup power must be considered. This becomes especially important in locations where hurricanes, seismic events, tornadoes, and ice storms are possible. However, with proper design and maintenance, a refrigeration system can be highly reliable.

3.2.3. Reactive Materials

Many inherently reactive materials are stored at chemical manufacturing facilities. Most commonly these are monomers for the production of plastics, fibre or elastomers. Under certain conditions, some normally stable materials may become reactive in the presence of a contaminant. To prevent spontaneous reaction in storage the materials are stored under conditions that eliminate, or almost eliminate, any reaction.

Common approaches to storing these materials include refrigeration to lower the reaction rate and reduce the possibility of the reaction being initiated, and addition of an inhibitor, most commonly used for monomers, that impedes the polymerization reaction. In many instances both of these strategies will be followed.

Where contaminants may cause a reaction, careful design is required to minimize the risk of contamination. Corrosion products are a common contaminant; selecting appropriate materials of construction is important in these cases. Where the contaminant is present in other parts of the process, routine analysis of feedstreams is important. If water can cause a reaction both design and operational measures may be required. Water may enter the process by many routes (utility connections, faulty tank roof drains, inadequately dried equipment, etc.); all of these must be either "designed out" or be very well managed via operating and maintenance procedures.

Lastly, as CFC refrigerant fluids are being phased out of refrigeration systems and other refrigerant fluids substituted, the reactivity of the substitute material with process materials must be evaluated.

3.3. Covers

3.3.1. Vapor Suppression Foams

While foam has traditionally been effective in the suppression of flammable and combustible liquid fires, its benefit as a suppressant of hazardous material vapors should not be underestimated. When used on spilled hazardous materials, a foam can insulate, absorb, suppress, and mitigate their reactivity (DiMaio and Norman, 1990).

A number of factors contribute to the effectiveness of foam as a vapor-suppressant. These include the type of foam, its expansion ratio, its drainage time, the rate of application of the foam (gal per min/ft^2), and its application density (gal/ft^2). Chemical foams have become obsolete, with mechanical foams now being used worldwide. A mechanical foam that has recognized attributes for vapor suppression is aqueous film-forming foam (AFFF). It is a synthetic foam (as compared to protein foams) with a surfactant that is part fluorochemical and part hydrocarbon. It suppresses vapors by forming an aqueous film produced by draining its foam bubbles.

Foams are categorized by their expansion ratio. Low-expansion foams have a ratio of 20:1 or less. The ratio of medium-expansion foam is 20:1 to 200:1 and for high-expansion foam it is 200:1 to 1000:1. AFFFs are typically low-expansion foams.

Another characteristic of foam is the concentration in which it is used. A 3% foam concentrate is mixed at a ratio of 97 parts water to 3 parts foam concentrate. Similarly, a 6% foam concentrate is mixed at a ratio of 94 parts water to 6 parts foam concentrate. Another key consideration is the 25% drainage time—the time required for 25% of the foam's liquid to drain from the foam. This characteristic can be used to measure foam stability, and in combination with the expansion ratio, it can be used to determine how the thickness of the foam blanket will vary with time (NFPA 471, 1989). This dictates the rate at which foam will have to be added to the initial blanket to ensure continued mitigation effects. Second, the drainage time can be used to determine the water, both total quantity and rate, that will have to be dealt with to prevent the incident from becoming an environmental incident as well.

When using foam as a vapor suppressant, the vapor pressure of the spilled material is important. That is to say, the higher the vapor pressure, the thicker the foam blanket should be.

Specific foams have been developed for use in vapor suppression and the reactivity mitigation of hazardous material spills. While they have been found to be effective on specific materials (Norman, 1987), such foams are no longer manufactured for commercial sale because of a lack of consumer demand. The foam presently recognized as today's prime vapor suppressant is a "universal" foam, a special type of AFFF with a biosynthesized polymer. Typically proportioned at 3%, this foam is appropriate for use on hydrocarbons, polar solvents, and other hazardous materials (Chubb National Foam, 1992a).

The "universal" foam has not been tested in the public arena to the extent that results are available in the literature. Proprietary testing by product users has shown that it is highly effective as a vapor suppressant when used at an expansion ratio of 35–45:1. At this ratio, its expansion was good and its drainage time was 13–15 minutes. This foam has been tested on the following materials with favorable results: trichlorosilane, dimethyldichlorosilane, phosphorus trichloride liquid, liquid hydrochloric acid, and titanium tetrachloride. Because of the quick vaporization rate of anhydrous ammonia and chlorine, no foam has been identified to date as being effective for all vapor suppression applications (Chubb National Foam, 1992b).

High-expansion foam can be used as an effective vapor suppressant on LNG spills. It adds heat from the water in the foam to the vapors as they pass through the blanket. This procedure induces buoyancy which can reduce downwind concentrations at ground level. Expansion ratios of 750:1 to 1000:1 have been found to be particularly effective (NFPA 11A, 1988). At the same time, a low-expansion foam in the 20:1 range would be effective for a hydrocarbon spill. As this shows, selection of the actual foam to be used must be based on the intended application and on the foam supplier's test data and recommendations.

When considering the application of foam to hazardous material spills, key requirements include an adequate water supply, the foam concentrate, appropriate proportioning equipment, piping, foam makers, and discharge devices and protective equipment for the personnel involved. Activation may be automatic, but in most cases it is manual. There are four types of application systems. They include fixed, semifixed, mobile, and portable (NFPA 11, 1988).

- **Fixed Systems:** These are complete installations piped from a central foam station, and discharging through fixed delivery devices into the area to be protected. Required pumps are permanently installed.
- **Semifixed Systems:** Two configurations for semifixed systems are possible. The piping to the area to be protected is put in place and the foam-making equipment must be connected to it when an incident occurs. In the second system, the foam is piped through an area from a central foam-making station and then the equipment to deliver the foam to the incident is connected. This equipment could consist of monitors, foam towers, hose lines, etc.
- **Mobile Systems:** A mobile system is any foam-producing unit that is mounted on wheels; either self-propelled or towed, that can be

brought to the incident. An available water supply for mixing the foam and the apparatus for delivering the foam are also required to complete the delivery system.

- **Portable Systems:** These are systems in which the foam-producing equipment and materials are transported by hand.

The periodic inspection and annual maintenance of foam systems and concentrate are key to their reliable operation. Full discharge tests can be conducted as part of personnel training activities that are key to effective and successful application.

3.3.2. Dry Chemical Covers

Dry chemical systems are well known for being an effective means of fire extinguishment of flammable and combustible liquid fires; however, certain types also have application as neutralizing and solidifying agents for liquid spills. Products are available for treating both acids and caustics.

Acids such as sulfuric, hydrochloric, nitric, phosphoric, perchloric, formic, acetic, chlorosulfonic, 50% hydrofluoric, and adipic can be treated by a mix of magnesium oxide and other chemical additives. It must be expected that when the dry chemical agent is applied to an acid spill, there will be a momentary increase in the volume of vapor coming off the spill. This puff is caused by the heat generated from the neutralization of the acid. To protect the personnel applying the dry chemical cover from this puff, suitable personnel protective equipment should be worn.

Such treatment results in a residue that is salt-based, e.g., magnesium sulfate from sulfuric acid. Thus, dry chemical agents should be readily available in areas where such materials are stored and handled, for example, tank truck transfer sites. The application systems or equipment for application of dry chemical agents should have their locations marked. These areas should be kept clear, marked, and inspected periodically in the same manner as fire extinguishers.

Acid-neutralizing and solidifying agents can be applied by shovel, hand-portable device, or larger wheeled or stationary units. A typical application measure might be a 20-lb pail of agent to a 20-ft^2 spill, which is a typical 5-gallon spill. Treatment time is usually 10 to 20 minutes, during which period only a moderate amount of heat is produced. The resulting residue will be pastelike and harden over time (ANSUL, 1991a).

Releases of caustics, such as sodium hydroxide, aniline, diethylamine, potassium hydroxide, and hydrazine, can be treated by a mix of citric and fumaric acid blended with other chemical additives. The treatment results in a residue that is salt-based (e.g., ammonium citrate from ammonium hydroxide). This type of agent should be stored and kept ready for use in areas where caustics are stored and transferred. The application equipment should be treated in the same manner as described for the treatment of acidic spills above.

As with the acid treatment, application can be made by shovel, hand-portable device, or larger wheeled or stationary units. A typical application measure might be a 10-lb pail of agent to a 10-ft^2 spill, which is a typical 3-gallon spill. Treatment time is usually 10 to 20 minutes, during which period only a moderate amount of heat is produced. The resulting residue will also be pastelike and harden over time (ANSUL, 1991b).

When disposing of the residue, determination of its residual pH is of critical importance, and therefore federal, state, and local regulations should be adhered to.

3.3.3. Other Covering Techniques

Ping-Pong Balls.   The use of hollow spheres of materials such as polyethylene or polypropylene, or other plastics that are spread over the surface of a liquid to reduce the rate of vaporization, has been studied. The rationale behind this technique is to cover the confined pool of liquid with lightweight balls that will float on the surface of the liquid and thus prevent vaporization.

Because the spheres used for the cover are light weight, they can easily be blown by the wind into one corner of the liquid pool they are supposed to cover. The larger the pool to be covered by ping-pong balls the more necessary it is to consider wind effects in designing the system.

In a test using liquid ammonia, the spill surface was covered with 0.8-inch diameter polypropylene balls to a depth of one ball. This resulted in covering 78% of the pool surface area (Feind, 1975). When compared to an equally uncovered pool, no difference in the rate of ammonia vaporization was observed. When the cover consisted of multiple layers of balls, a reduction in vaporization was effected.

Water.   For substances that have a density greater than water and are either insoluble in it, or only slightly soluble in it, water can be an effective cover in the event of a release—as long as the materials are confined in a diked area. Examples of two such materials in this category would be liquid bromine and carbon disulfide.

Water must be added to the top of the spilled material slowly, using a fine spray or low-velocity nozzles located along the side of the diked area.

Breakup of the spilled pool of material is prevented when the water addition velocities are kept within the laminar flow regime.

In some cases where water is used as a cover for an acidic material, the addition of a base, such as sodium bicarbonate or ammonia hydroxide, in low concentrations has been suggested. In such cases, the effects of the heat of reaction from the neutralization and liberation of carbon dioxide from use of the carbonate must be taken into consideration.

Adsorbents. The use of dry chemicals as adsorbents has been discussed above. However, there are other adsorbent materials, such as activated charcoal, hay, corn cobs, sawdust, and such, that can be utilized as well. Most of them would work on many organic compounds as long as the spill was confined to a small area by either dikes or curbs, or as long as the spill was small and covered only a minimal area. The major problem in this application technique is that the personnel who apply the materials must wear adequate protective gear. An alternate technique would be to blow the absorbent material over the spill from a distance, using suitable machinery, such as a snow blower or forage blower unit (Greer, 1976).

Other Covers. Other materials that can be utilized to cover the surface of the spill and stop vaporization are sheets of plastic, such as polyvinyl chloride, polyethylene, polypropylene, or polystyrene. The effectiveness of this approach is dependent on the size of the covering material available and the area to be covered. As with the other techniques described earlier, using a cover as a mitigation measure for a spill requires prepositioning and preplanning. Prepositioning is necessary so the cover will be close to the spill and readily available to those responding to the incident. Preplanning calls for envisioning the scenarios in which the cover would be used and then training response personnel in its application. When conducting drills, the response team should also be wearing the protective equipment they would wear in an actual incident. The potential of some of the plastic covering materials to produce static sparks must be considered carefully, especially if the substance to be covered is flammable or combustible.

Wet sand can also be used as a covering material, to deal with a leak or spill of elemental phosphorus. In this situation the sand acts as a cover to keep the phosphorus away from the oxygen in the air while the water reacts with it in a more controlled manner. Again, in this application, the provision of proper protective equipment for the personnel applying the wet sand is essential.

3.4. Deliberate Ignition

Deliberate ignition of a release in which the material is both toxic and flammable is a means of mitigation. But this approach may only be appropriate in a few select material-specific and facility-specific cases. Certain materials may well represent a toxicity hazard greater than the flammability hazard. Moreover, the combustion of the vapor cloud may be incomplete or generate toxic products (Husa and Bulkey, 1965). The facility itself, the layout of the processes, other hazardous materials, property lines, and prevailing weather conditions all are of critical importance when a vapor cloud is ignited.

Timely ignition is critical, so there is a need for immediate or continual activation of the system being used. Devices that can be used for remote ignition include flare pistols, electric ignition devices, and propane burners. The location and spacing of the devices are typically functions of the material and facility; however, it is critical that a release does not go undetected by the gas detectors (if there is no continual ignition source) or by the ignition system itself.

Deliberate ignition is a mitigation means that introduces risk. Its viability should be weighed carefully, along with the risks presented by the potential release, in deciding whether or not deliberate ignition can be employed. Detailed analyses and tests should be conducted prior to relying on deliberate ignition as a proven means of minimizing risk both onsite and offsite.

A related issue is extinguishing an existing fire before the release can be controlled. An example is a burning sour gas wellhead blowout. When ignited, the toxic hazard is sulfur dioxide, which is carried aloft in a thermally buoyant plume. Once extinguished, the toxic hazard is constituted by the presence of hydrogen sulfide which, with certain gas compositions, may form a toxic heavy vapor cloud. This issue needs to be considered during emergency preplanning.

3.5. References

American Industrial Hygiene Association. 1992. *Emergency Response Planning Guidelines for Air Contaminants.* Fairfax, VA: American Industrial Hygiene Association, May.

ANSUL. 1991a. Spill-X-A Acid Neutralizer Data Sheet. Marinette, Wisconsin: ANSUL Fire Protection.

ANSUL. 1991b. Spill-X-C Caustic Neutralizer Data Sheet. Marinette, Wisconsin: ANSUL Fire Protection.

CCPS (Center for Chemical Process Safety) 1988a. *Guidelines for Vapor Release Mitigation*. New York: American Institute of Chemical Engineers.

CCPS (Center for Chemical Process Safety) 1988b. *Guidelines for Safe Storage and Handling of High Toxic Hazard Materials*. New York: American Institute of Chemical Engineers.

CCPS (Center for Chemical Process Safety). 1989a. *Guidelines for Chemical Process Quantitative Risk Analysis*. New York: American Institute of Chemical Engineers.

CCPS (Center for Chemical Process Safety). 1989b. *Guidelines for Technical Management of Chemical Process Safety*. New York: American Institute of Chemical Engineers.

Chubb National Foam. 1992a. Universal Gold 3% Data Sheet. Exton, Pennsylvania: Chubb National Foam.

Chubb National Foam. 1992b. Personal communication with Arthur D. Little, Inc. Cambridge, Massachusetts.

DHHS (NIOSH) (Department of Health and Human Services, National Institute of Occupational Safety and Health). 1990. *NIOSH Pocket Guide to Chemical Hazards*. U.S. Department of Health and Human Services, Center for Disease Control, Cincinnati, Ohio.

DiMaio, L. R., and E. C. Norman. 1990. Continuing Studies of Hazardous Material Vapor Mitigation Using Aqueous Foams. *Plant/Operations Progress*, 5(3): 135.

Feind K. 1975. Reducing Vapor Loss in Ammonia Tank Spills. *Safety in Ammonia Plants and Related Facilities Symposium, 17th Annual Proceedings*, pp. 114–118. New York:. American Institute of Chemical Engineers.

Greer, J. S. 1976. Feasibility Study of Response Techniques for Discharges of Hazardous Chemicals That Float on Water. Report No. CG-D-56-77. U. S. Washington, D.C.: Department of Transportation, October.

Greer, J. S., and S. S. Gross. 1980. The Practicality of Controlling Vapor Released from Spills of Volatile Chemicals through Cooling Control of Hazardous Material Spills. *Proceedings of National Conference*, pp. 130–133. Nashville, TN: Vanderbilt University.

Husa, H. W. and W. L. Bulkey. 1965. Hazards of Liquid Ammonia Spills. *Safety in Air and Ammonia Plants*, 7, no. 42.

Mackay, D., and R. Matsugu. 1973. Evaporation rates of hydrocarbon spills on water and land. *Canadian Journal of Chemical Engineering*, 5: 434.

Melhem, G. A., and P. A. Croce. 1994. "Advanced Consequence Analysis: Emission, Dispersion, Fires and Explosion Dynamics," Working Manuscript, Arthur D. Little, Inc.

NFPA 11. 1988 Ed. Standard for Low Expansion Foam and Combined Agent Systems. Chapter 2. Quincy, Massachusetts: National Fire Protection Association.

NFPA 11A. 1988 Ed. Standard for Medium- and High-Expansion Foam Systems, Appendix A. Quincy, Massachusetts: National Fire Protection Association.

NFPA 471. 1989 Ed. Recommended Practice for Responding to Hazardous Material Incidents. Appendix A-6-4.1.9. Quincy, Massachusetts: National Fire Protection Association.

Norman, E. C. 1987. A Guide to the Use of Foam on Hazardous Material Spills. *Hazardous Materials and Waste Management Magazine*, Sept/Oct.

4

Fluid Curtains

4.1. Introduction

The main objective of a fluid curtain is to help mitigate explosions by absorbing energy as droplets break up and by creating an increased total surface area to reduce potential flammable and toxic hazard zones by dilution with air or by chemical reaction with the water or reactive materials contained in the curtain.

Dilution is attributed to the large amounts of air that are entrained by the spray. As a result of air entrainment, dispersion behavior is altered for materials that exhibit negative buoyancy upon release. This is found to be effective in controlling flammability hazards that are located close to the release source. If an explosion does occur, some of the energy will be absorbed in the breakup of the water spray droplets, thereby mitigating the explosions impact.

Hazard-zone reduction can also be achieved by reducing the source emission strength using mass transfer, that is, chemical reaction and/or absorption. While dilution is effective in reducing near-field hazards, reducing source strength can be effective in reducing far-field hazards caused by toxic materials.

4.2. Previous Work

Table 4.1 summarizes research work on fluid curtains dating back to the 1970s. The research includes both small- and large-scale experimental work, as well as theoretical analysis. While the published data are very useful in furthering our understanding of the effectiveness of fluid curtains as a mitigating measure, and their parametric dependence, they do not provide a precise method for design. In the following sections, we examine absorption and dilution by fluid curtains, describe simple models for absorption and dilution, and provide guidance on their applications for design.

TABLE 4.1
Summary of Fluid Curtain Research

Author	Year	Objective of Study	Major Findings and Activities
Briffa and Dumbrowski	1966	To study the machanism of air entrainment into a flat (fan) spray nozzle.	• The mass of air entrained into the spray, decay of air velocity along the spray axis, and the spread of drops in the plane normal to that of the spray sheet were related to operating conditions by theoretical equations confirmed by the experimental work
Eggleston, Herrera, and Pish	1976	To provide needed data about the use of air entrained by a water spray to dilute flammable vapor releases below the lower flammability limit.	• Absorption/adsorption effects are insignificant in the case of ethylene and vinyl chloride. • Sprinklers and water-spray nozzles vary widely in their efficiency as air movers. • Flame quenching was not affected in any of the experiments • Water sprays increased the rate of flame propagation. • The air-pumping action of a water curtain can be used to set up a barrier to the horizontal flow of vapors.
Heskestad, Kung, and Todtenkopf	1976	To study the entrainment of air into water sprays discharging into quiescent surrounding, both theoretically and experimentally	• Developed a simple theoretical model for air entrainment flow in a water spray discharging downward in quiescent space. • Air entrainment flow is sensitive to water discharge rate when nozzle pressure is varied at constant nozzle diameter. • Entrainment flow is insensitive to discharge rate when nozzle diameter is varied at constant nozzle pressure.
Martinsen, Muhlenkamp, and Olson	1977	To test varied spray nozzles and water flow rates to determine their effectiveness in dispersing LNG vapors.	• Water sprays do affect concentration reduction of LNG vapors

TABLE 4.1 (continued)

Author	Year	Objective of Study	Major Findings and Activities
Moore and Rees	1981	To force dispersion by water and steam. Eight experiments were conducted.	• Best water-spray configuration is upward, with conical pattern and narrow angle. • Best steam-jet configuration is downward. • Water is more effective downstream in reducing 15-m downwind concentrations.
Van Doorn	1981	To investigate the use of water sprays in dispersing a hazardous cloud by using air entrained by the spray to dilute the material to safe limits.	• A semiemperical model was developed of describe transfer of momentum between spray curtain and air. • Induced air movement is turbulent and strongly dependent on initial water velocity. • Nozzle design and operating parameters influence spray-curtain performance. • When two spray-curtains are used, the one with the larger flow predominates.
McQuaid and Fitzpatrick	1981	To determine the response of a heavy-gas plume to the action of a water spray barrier.	• A model was developed that was applicable to water-spray barriers.
Moodie	1981	To study the effectiveness of different nozzle sizes and arrangements in full-scale water-spray barriers to disperse heavy-gas vapor clouds.	• Amount of air entrained by a water-spray barrier may be dependent on nozzle inclination. • Results demonstrated the effectiveness of the basic arrangement tested and the experimental method of assessment.
Meroney and Neff	1984	To calculate the behavior of heavy and cold gas clouds subject to dilution by a water spray curtain, using a numerical model. Work included model validation.	• Variations in curtain position, spray entrainment velocity, wind speed and source conditions resulted in a wide range of spray-curtain performances.

TABLE 4.1 (continued)

Author	Year	Objective of Study	Major Findings and Activities
Moodie	1985	To gather experimental data to allow the design and prediction of the performance of water-spray barriers.	• Showed the best water-spray orientation is pointing upward. • Increasing the specific momentum flow rate improves barrier effectiveness at a given wind speed. • Savings in water usage are possible by selecting nozzles based on both momentum flow number and flow number characteristics.
Emblem and Madsen	1986	To check the efficiency of a full-scale water curtain installation by simulating an actual gas release using CO_2 as the test gas.	• The main effect of the water curtain was to dilute the maximum concentration region and form a cloud of almost homogeneous concentration. • In this case the nozzle spacing was so wide that the effect of the barrier was small.
Blewitt, Yohn, Koopman, Brown, and Hague	1987	Test the effectiveness of water sprays to mitigate HF releases.	• Water curtain spray systems achieved approximately a 36 to 49 percent reduction in downwind concentrations of HF. This reduction is consistent with labortory experimental data. • The compressed air and water spray appeared to be less effective than the water-spray curtain.
Schatz and Koopman	1990	Determine the effectiveness of water sprays for mitigating releases of HF in small flow chamber tests. Determine the effectiveness of water sprays for mitigating HF and alkylation unit acid (AUA) in large-scale field trials.	• Tests showed that releases of HF and AUA can be successfully mitigated by properly designed and applied water sprays from spray curtains or monitors. Field tests showed efficiences of >90% at water-to-HF ratios of 40:1 or higher. • The effectiveness of water sprays in mitigating acid releases as a function of flow conditions, water, acid, air properties and geometric factors was accomplished. • Water to hydrofluoric acid ratio was found to be the major variable.
Van Zele and Diener	1990	To investigate the effectiveness of water sprays in reducing hazards from HF releases. Many key variables were studied to enhance HF removal efficiency.	• Water-to-HF ratio is key. • Upflow water sprays are more efficient than downflow sprays. • Removal efficiency depends on spray nozzle configuration, nozzle size, spray pattern, spacing, etc.

TABLE 4.1 (continued)

Author	Year	Objective of Study	Major Findings and Activities
Fthenakis, Schatz, and Zakkay	1991	Develop computer simulationss of the Hawk, Nevada Test Site, field tests of water sprays to absorb HF releases.	• Computer model HFSPRAY developed amd validated against all the Hawk field data. • Accurate specification of nozzle parameters is essential; sensitivity analysis with model indicated that not only drop size variations but also variation of initial drop velocity and spray pattern can have a measurable effect on model predictions.
Blewitt, Peterson, Ratcliff, and Heskestad	1991	Collect information needed to aid in the design of a water-spray mitigation system for an industrial facility by using a scale model in an open-circuit boundary layer wind tunnel.	• Nozzles pointed up and down at a 45° angle appeared to be less effective than nozzles directed inward in incereasing cloud spreading and retarding cloud penetration of the spray curtain. • Changing the nozzle spacing from 0.5 m to 0.9 m spacing had little effect on cloud behavior. • The 55-m-long water spray config-uration with nozzles pointed hori-zontally into the wind appeared to be effective in increasing cloud spreading and retarding cloud pene-tration of the spray curtain. This con-figuration did not allow a surface release (directed with the wind) to penetrate beneath the curtain. The disadvantage of this configuration is that cloud material can be deflected around the edges of the spray manifold. • A 100-m U-manifold with 1.83-m nozzle spacing was effective in increasing cloud spreading and retarding cloud penetration with horizontal or up/down sprays of low wind speeds (3 m/s). • For releases directed against the wind under low speeds, a complete enclosure appeared necessary to ensure water-spray contact with the vapor cloud.

TABLE 4.1 (continued)

Author	Year	Objective of Study	Major Findings and Activities
Petersen and Blewitt	1992	Study effectiveness of water sprays and fire monitors in mitigating HF releases with varying wind speeds and directions in two different refinery unit configurations.	• For the refinery unit that had many low level obstructions, it was found that the placement and aiming of spray nozzles was difficult. As a result, fire monitors appeared to give the best overall performance due to the complexity of the structures around the source of the leak. • For the large refinery unit, spray placement and aiming was easier to do. In this case, the water-spray curtain was found effective in delivering water to the cloud.
Fthenakis and Blewitt	1993	Assess the effectiveness of two water spray configurations in mitigating an accidental release of HF in an actual facility.	• The performance of an HF mitigation system comprising a series of spray nozzles was assessed by computer simulation with the aid of wind tunnel experiments. According to the simulations, a two-tier horizontal configuration removed HF more effectively than the up-and-down configuration, for the specific flow rates, release heights, and wind speeds considered. • For high wind speeds (17 m/s), a 70% effectiveness was found. • For average wind speeds (5 m/s), a 96% effectiveness was found.

4.3. Absorption/Mass Transfer

Absorption by water sprays or by sprays using water and a selective reactant for a specific chemical can be an effective method of reducing the concentration of a released material. For example, ammonia, hydrogen fluoride, hydrogen chloride, and hydrogen cyanide are highly soluble in water. On the other hand, chlorine, sulfur dioxide, hydrogen sulfide, and nitrogen oxide are only slightly soluble in water. Ammonia undergoes a fast irreversible ionization reaction when it comes in contact with water:

$$NH_3 + H_2O \rightarrow NH_4OH$$

$$NH_4OH \leftrightarrow NH_4^+ + OH^-$$

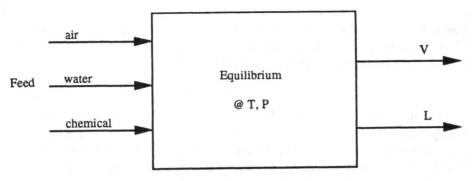

Figure 4.1. An equilibrium stage.

Prugh (1986) derived a theoretical model to assess the effectiveness of water sprays based on flow rates of released chemical, air, water, and the assumption of physical equilibrium at a constant temperature and pressure. Consider the volume element illustrated in Figure 4.1.

Three streams enter the volume, including air, water, and the chemical substance whose hazard we are trying to control. When n is defined as the number of moles, y is defined as the mole fraction in the vapor phase, and x is defined as the mole fraction in the liquid phase. The subscripts C, a, and w correspond to the chemical in question, air, and water, respectively. The individual molar flow rates of these three species are

$$\frac{dn_C}{dt} = \frac{1}{m_{w_C}} \frac{dm_C}{dt} \qquad (4.1)$$

$$\frac{dn_a}{dt} = \left[\frac{C_o - C_C}{C_C} \right] \frac{dn_C}{dt} \qquad (4.2)$$

$$\frac{dn_w}{dt} = \frac{dm_w}{dt} \frac{f}{18} + H_w \frac{R_H}{100} \frac{dn_a}{dt} \qquad (4.3)$$

where m is mass in kilograms, t is time in seconds, H is Henry's law constant, C_o is the initial concentration, usually 100%, C_C is an average crosswind concentration of the hazardous chemical in volume percent, R_H is the relative humidity, and f is the fraction of the water flow that contacts the vapor cloud (<0.2). Henry's law constant is defined (Carroll, 1991) as

$$H_{ij} = \text{limit} \frac{f_i^L}{x_i} \qquad \text{as } x_i \to 0 \qquad (4.4)$$

where ij indicates that H is for solute i in solvent j, and f_iL is the liquid phase fugacity. If we assume ideal behavior in both the liquid and vapor phases, the requirement of physical equilibrium is

$$x_iH_{ij} = y_iP \tag{4.5}$$

Equation (4.5) is probably the form that first comes to mind when Henry's law is mentioned. It is applicable up to two atmospheres of pressure (P) and liquid mole fractions (x_i) of up to 1%. Higher values of pressure or liquid mole fractions require corrections using vapor phase fugacity coefficients and liquid activity coefficients or fugacity coefficients.

An overall mass balance around the volume of Figure 4.1 yields

$$F = \frac{dn_a}{dt} + \frac{dn_C}{dt} + \frac{dn_w}{dt} \tag{4.6}$$

or

$$F = L + v \tag{4.7}$$

where F is the total molar flow rate in kmol/s, L is the equilibrium liquid molar flow rate, and v is the equilibrium molar vapor flow rate. An individual mass balance on the chemical and water yields

$$z_CF = x_CL + y_Cv \tag{4.8}$$

and

$$z_wF = x_wL + y_wv \tag{4.9}$$

where z is the overall feed mole fraction:

$$z_C = \frac{dn_C/dt}{dn_a/dt + dn_C/dt + dn_w/dt} \tag{4.10}$$

and

$$z_w = \frac{dn_w/dt}{dn_a/dt + dn_C/dt + dn_w/dt} \tag{4.11}$$

Physical equilibrium is satisfied when

$$y_C = K_Cx_C + H_Cx_C \tag{4.12}$$

and

$$z_w = K_wx_w + H_wx_w \tag{4.13}$$

Substituting y by xH and v by $F - L$ and ensuring that the liquid mole fractions sum to 1 yields equation (4.14) which can be solved iteratively for L/F:

$$x_C + x_w = \frac{Z_C}{[1 - L/F]H_C + L/F} + \frac{Z_w}{[1 - L/F]H_w + L/F} = 1 \quad (4.14)$$

Once the value L/F is established, the value of y_C can be calculated from $x_C H_C$. The effectiveness of the spray curtain, or dilution factor, FD is

$$FD = \frac{C_C/100}{y_C} \quad (4.15)$$

The approach followed by Prugh (1986) is useful in estimating the overall effectiveness of a specific water-spray curtain, but important parameters such as drop size and gas–liquid mass transfer are not accounted for.

The effectiveness of absorption depends on water flow rate, drop size, and gas solubility. Fthenakis (1989) developed a simple model for evaluating mass transfer induced by a single water-spray nozzle for a point-source release:

$$\frac{C(x)}{C(0)} = \exp\left[-\frac{K_g a}{G(1 + K_g amz/2L)} x\right] \quad (4.16)$$

He considers a nozzle located at an elevation z above the ground, spraying water at a rate of L gmoles/m^2/s over an area of A m^2. All water drops have the same diameter d and velocity u. If the source strength of the point source inside the spray cone is G gmoles/m^2/s, and assuming that the gas is well mixed in the vertical direction, the gas phase concentration of a given chemical can be calculated as a function of radial distance within the spray. In equation (4.16), a is the interfacial spray area per unit volume, K_g is the mass transfer coefficient, x is the axial distance from the spray cone center, m is the equilibrium curve slope from equation (4.19), and a is given in equation (4.17) as

$$a = \frac{1}{\rho_w} \frac{dM_w}{dt} \frac{6t}{dV} \quad (4.17)$$

where V is the spray volume in m^3 and t is the drop exposure time in seconds, which depends on the ratio of elevation-to-drop velocity:

$$t = 1.40 \frac{z}{u} \quad (4.18)$$

The equilibrium curve slope m can be estimated from Henry's law.

$$m = C \exp[kT] \qquad (4.19)$$

where T is the water temperature in Celsius units and C and k are constants. For example, the equation for ammonia is

$$m = 0.36 \exp[0.053T] \qquad 0 < T < 38 \qquad (4.20)$$

Fthenakis and Schatz (1991) and Fthenakis and Zakkay (1990) present a more complex analysis of absorption by water sprays. A two-dimensional model using the conservation laws of mass, momentum, species, and energy incorporated the k-epsilon turbulence model to describe the gas phase. The liquid phase was modeled using a Lagrangian approach for a finite number of particles. Model results were compared against the Hawk HF data and appear to be in good agreement. Trends predicted by the model indicated increased hydrogen fluoride (HF) removal efficiency with decreasing water drop size and decreasing wind speed. Modeling results also indicated that sprays used in upflow were more effective than those used in downflow for the same water flow rate.

Experiments with hydrogen fluoride (HF) involving absorption and chemical reaction by water are summarized by Blewitt et al. (1987). These experiments involved open-field and laboratory-scale tests. In the open-field experiments, liquid hydrogen fluoride was released and controlled by water sprays located 35 m from the source. Plume widths of HF observed at 35 m ranged from 15 to 20 m. Wind speeds ranged between 5.4 and 6.8 m/s and the atmospheric stability was neutral. Two spray headers, consisting of 25 nozzles each, with a nozzle spacing of 1.524 m, were tested. The first header was directed from ground level. The second was pointed downward from an elevation of 3.658 m. The spray systems delivered a total water flow rate of approximately 2.65 m³/min., yielding a 21:1 ratio of water to HF on a mass flow basis. Fog-type nozzles produced an average droplet size of 250 microns. The removal efficiencies reported ranged from 36% to 49% of the HF released. Removal efficiency was established from samplings taken at 300 m from the source as well as liquid collection. It is important to note that these efficiencies should be attributed to absorption alone. Dilution effects at the source produced little concentration reduction in the far field.

A summary of the laboratory data is presented in Table 4.2. The data indicate that removal efficiencies increased with increasing water-to-HF mass ratios.

TABLE 4.2
Effects of Water Rates on HF Removal Efficiencies
(Blewitt et al., 1987)

Water/HF mass ratio	HF Removal Efficiency (%)
7	8.7
10	37.8
19	44.6
22	54.7
43	69.7
64	78.8

Schatz and Koopman (1990) reported on the Hawk series, 87 tests conducted at the DOE Nevada test site. These experiments were large-scale chamber releases of HF, as well as laboratory experiments. The objective of these tests was to study the effect of the water-to-HF ratio, water spray geometry, water application via a fire monitor, acid type (anhydrous HF and alkylation unit acid (AUA)), acid temperature and pressure, water additives, relative humidity, wind speed, and steam as an acid jet dispersant on HF removal efficiency. Figure 4.2 shows removal efficiencies ranging from 25 to 90% for water-to-HF volumetric flow ratios ranging from 6:1 to 40:1. Fire monitors provided removal efficiencies comparable to those of water sprays. Some of the conclusions reached by the authors were:

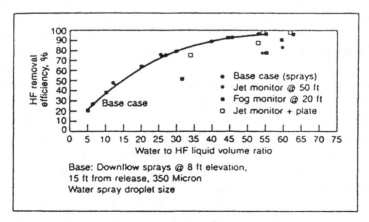

Figure 4.2. Effects of water rates on HF removal efficiencies (Schatz and Koopman, 1990).

- The water-to-HF mass ratio is a key variable.
- Water additives and steam as a jet dispersant had little effect on the removal efficiency for HF or AUA (alkylation unit acid).
- Removal efficiency increased with decreasing droplet size, decreasing spray header elevation (4.87 vs. 2.43 meters), and increasing distance between spray nozzles from 0.30 to 1 meter.
- Upflow water sprays provided better removal efficiency than downflow water sprays.
- The type of acid HF or AUA had a negligible impact on removal efficiency.
- Relative humidity and wind speed were found to have negligible effects on removal efficiencies for HF or AUA under the conditions tested.

4.4. Air Dilution

Water sprays induce dilution by entraining large volumes of air. Air entrainment, which is induced by momentum transfer from the water droplets to the ambient air, is a strong function of spray configuration, droplet size, droplet velocity, and spray location. Large droplets entrain less air than smaller droplets but induce better mixing and dispersion. Although smaller droplets entrain more air, they have smaller terminal velocities which are almost constant and cause poor mixing and dilution. Alternatively, when chemical reaction is involved, smaller droplets result in longer contact time and a larger surface-to-volume ratio, which enhances mass transfer.

Water curtains or screens can also act as a physical barrier to the downwind travel of vapor clouds. Figure 4.3 illustrates the flow field of the spray curtain. The outflowing air at ground level can be utilized to disperse and divert gas clouds from potential ignition sources. This is achieved by creating a wall jet that spreads outward along the impact surface, that is, the ground. However, this may require a large volumetric air flow with large velocities, that is, spray nozzles with large flow numbers.

The impingement region illustrated in Figure 4.3 starts at values of z/H ranging from 0.75 to 0.85, as determined by Watts (1976) and Gutmark et al. (1978), where z represents a height above the impingement surface and H is the height of the spray nozzle above the impingement surface.

Donaldson and Snedickar (1971) studied jet impingement on flat plates as well as on convex- and concave-shaped objects. Base pressure changes were measured for all considered geometries and velocity measurements reported for the flat-plate geometry. The study indicated that wall jets are

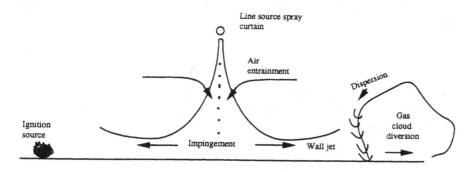

Figure 4.3. A line source spray curtain pointing downward .

formed with self-similar velocity profiles in all azimuthal directions for normal impingement. Well-formed wall jets were not observed for flat-plate locations beyond 25 times the jet nozzle diameter. Looney and Walsh (1984) provided a comprehensive review of research on free and impinging jet flows. Looney and Walsh (1984) also showed that the spreading rate of wall jets on impact is a constant, approximately equal to 0.1 m/m.

A good estimate of air entrainment flow in a spray can be obtained by examining in detail the momentum exchange between the spray and the surrounding air. We present a similar model to the one developed by Heskestad et al. (1976). In deriving this simple model we assumed the following:

- The spray is pointing downward.
- The effects of droplet motion on drag is negligible.
- Spray droplet size distribution is uniform.
- Air velocity in a spray cross-section is uniform.
- The coalescence of droplets is negligible.
- The aerodynamic drag on a given droplet is not influenced by its neighboring droplets.

The forces acting on a single droplet include gravity, which acts vertically downward, and drag, which depends on the droplet shape and its relative velocity with respect to surrounding air. A force balance on a single droplet yields

$$m_d \frac{du_d}{dt} = m_d g - F_d$$

(4.21)

where m_d is the droplet mass, F_d is the aerodynamic drag on the droplet, g is acceleration due to gravity, and u_d is the drop velocity.

Ignoring the time dependence of droplet velocity, we can rewrite equation (4.21) as

$$\frac{du_d}{dz} = \frac{g}{u_d} - \frac{F_d}{m_d u_d} \tag{4.22}$$

where z is the vertical distance from the droplet source.

The increase in momentum flux in the air phase is attributed to the force exerted by the liquid droplets:

$$\frac{d}{dz}[\rho_a u_a^2 A] = \frac{F_d}{u_d}\left[\frac{Q_w \rho_w}{m_d}\right] \tag{4.23}$$

The term in brackets on the right-hand side of equation (4.23) is the number of liquid droplets, where F_d is the drag force per droplet shown in equation (4.24), Q_w is the volumetric flow rate of water, ρ_w is the density of water, and ρ_a is the density of air.

The drag force per droplet (F_d) is expressed as

$$F_d = C_D A_d \left[\frac{\rho_a (u_d - u_a)}{2}\right]^2 \tag{4.24}$$

where A_d is the droplet frontal area and C_D is the drag coefficient, which is a function of the droplet's Reynolds number (N_{Re}).

The Reynold's number N_{Re} is based on the relative velocity $u_{d,a} = u_d - u_a$ between the droplet and the surrounding medium:

$$N_{Re} = d_d \, |u_{d,a}| \, \rho_a \tag{4.25}$$

The drag coefficient C_D in equation (4.24) is related to the droplet Reynolds number by equations (4.26) to (4.29).

$$C_D = \frac{24}{N_{Re}}, \qquad\qquad\qquad\qquad\qquad N_{RE} < 0.1 \tag{4.26}$$

$$C_D = \frac{24}{N_{Re}}\left[1 + \frac{3}{16}N_{Re} + \frac{9}{160}N_{Re}^2 \, [\ln(2N_{Re})]\right], \quad 0.1 < N_{Re} \le 2 \tag{4.27}$$

$$C_D = \frac{24}{N_{Re}}[1 + 0.15 N_{Re}^{0.687}], \qquad\qquad\quad 2 < N_{Re} < 500 \tag{4.28}$$

$$C_D = 0.44, \qquad\qquad\qquad\qquad 500 < N_{Re} < 200{,}000 \tag{4.29}$$

Equations (4.21) and (4.23) can be integrated numerically to provide an estimate of the air entrainment rate into the spray. Heskestad et al. (1976)

compared their model's prediction to experimental data and found significant agreement. Entrainment rates were predicted to within 17% of the measured ones. The initial conditions for integration used by Heskestad et al. (1976) for droplet velocity are

$$u_d(z_0) = C_M \frac{Q_w}{A} \tag{4.30}$$

where C_M is the momentum coefficient of the nozzle. Heskestad et al. (1976) found an experimental value of 0.41 for water pressures in the range of 97 to 143 kPa. The value of C_M can be estimated for wide-angle sprays using the following equation:

$$C_M = \frac{2}{\tan^2(\theta/2)} \left[\sqrt{1 + \tan^2(\theta/2)} - 1 \right] \tag{4.31}$$

where θ is the spray angle. This expression agrees with the single experimental data point for C_M (Heskestad et al., 1976).

The analysis for the single spray can be extended to curtains by using an aggregate surface area and water flow rate. An expression relating A, the spray area, to z, the axial distance, is required. This simple theoretical model can be modified to predict the spray envelope as well.

4.5. Defining Spray Requirements for Mitigation

Water sprays are simple, reliable, and inexpensive and can be either fixed or mobile. However, their effective use for protection requires careful evaluation of credible release rates, release durations, chemical characteristics, spray hydrodynamics, and mass transfer. The efficiency of a water-spray system is a function of the following:

- whether the spray is fixed or mobile
- distance from the release point;
- number of nozzles;
- nozzle position/spacing;
- nozzle type;
- nozzle orientation;
- nozzle source pressure;
- flow rate of water and duration of its release.

Before designing a spray system to mitigate spills of toxic and/or flammable materials, design requirements should be defined. One method of doing this is to:

1. Define potential scenario, chemical properties, source geometry, and atmospheric conditions, which can affect the width of the cloud, the effect of obstacles on the cloud and the impact of the water sprays themselves;
2. Calculate hazard zones using well-established and validated hazard models, using experimental data or previous incident data when available;
3. Determine dilution and/or absorption factors that are required to ensure reduction of the hazard zone to an acceptable level;
4. From the results of step 3, determine the spray characteristics required using the simple models outlined in this chapter to assess the design sensitivity to liquid droplet size, spray area, solubility, contact time, orientation, wind speed, etc.; and
5. Define nozzle type, number, position, orientation, liquid flow rates, source pressure, etc.

Keep in mind that it is difficult to design a system that will allow quick and effective activation for all possible ranges of meteorological conditions, and that better spray effectiveness is obtained for materials that are highly soluble in water. While effective for small spills, dilution alone is not practical for large spills. The effects of dilution can be significant in reducing hazard zones in the near field close to the source. The effect of dilution on reducing concentrations at large downwind distances, however, is small.

The time required for actuation is a critical issue. Cloud travel time to sensitive areas can be achieved in a few seconds for jet releases and in a few minutes for refrigerated releases. Therefore, to be effective, water curtains must be designed to be activated within a short time of the onset of a release.

Water curtains can be an effective postmitigation technique. Their optimal design is best addressed on a case-by-case basis.

Important design aspects of fixed water sprays include release orientation, release momentum, and release contact with the water sprays. If the release is a two-phase jet and has a momentum that is larger than that of the water spray, the jet will penetrate the water spray with little interaction which will lead to a poor removal efficiency. Water-spray efficiency can be maximized by ensuring that all of the water comes in contact with the vapor cloud, that is, by ensuring that the cloud will spread out across the entire area covered by the water spray such that all the water is used.

4.5.1. Water Curtain Design Example

This example examines the design of a water-spray system that is used for the mitigation of anhydrous ammonia in the event of an accidental release. The NH_3 tank is capable of storing 35 m^3, it is 11 m in length and is cylindrical in shape lying in the horizontal plane. The base of the tank is surrounded by a protective dike or a bund. All piping connections are at the bottom of the tank. This is shown in Figure 4.4.

It is important to note the special design aspects of this example. Metal detection plates are installed on all sides to prevent any momentum-dominated two-phase release from going through the water spray. A two-phase release will impact the dike walls or deflection plates, causing the liquid ammonia to coalesce and rain out inside the dike, so that only vapor (without momentum) will come in contact with the water spray through the opening between the deflection plates and the tank.

One credible leak scenario is a two-phase discharge of ammonia from the tank (possibly due to a hole in the tank due to corrosion). The presence of the dike and the deflection plates will reduce the release momentum and force the liquid to collect in the dike. Assuming that 3.8 kg/s is airborne upon release (the unmitigated scenario), a downwind distance to 1000 ppm was modeled to be 2 km, assuming neutral buoyancy, as shown in Table 4.3.

The spray system characteristics are illustrated in Figure 4.4. Water flow is uniform through all 20 nozzles exiting at a 140° angle, producing a fine mist barrier that is conical in shape. Wind velocity is assumed to be 2 m/s. Table 4.3 shows a summary of predicted downwind distances for the unmitigated and mitigated scenarios.

The water spray will be designed to reduce the downwind distance to less than 0.2 km to 1000 ppm. Using the simple model outlined in the previous section, we estimate a water requirement of 300 GPM or 19 kg/s. The water curtain created by the spray nozzles can be expected to mitigate vapor releases by two means. The first is through dilution which is experienced when the water spray entrains air into its flow and enhances the rate of mixing between the NH_3 vapor and the air. The second would be through adsorption. Because of the high affinity of ammonia for water, much of the vapor could be expected to be adsorbed into the water spray. At 0°C 89.9 grams of NH_3 are soluble in 100 grams H_2O.

This example is presented for illustrative purposes only, and is designed to demonstate the effect of maximum removal efficiency. Turbulent mixing, which can reduce the actual removal efficiency, has not been included in this example. Procedures to account for turbulent mixing, jet momentum

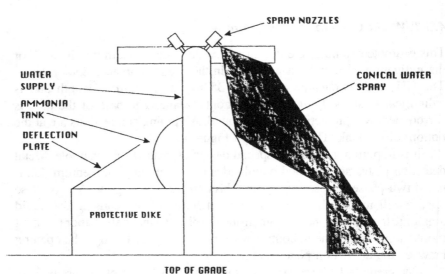

TANK DETAILS
34.822 m³ capacity
Liquid Level at 805
Length = 11.05 m
Length with dike = 12.8 m
Width with dike = 4.42 m
Maximum leak rate = 3.8 kg/s
Storage Pressure = 1000.4 kPa
Storage Temperature = 25°C

NOZZLE DETAILS
Diking Model C-2 (20 nozzles)
140° angle
3.9 K Factor or 0.000946 m³/s/nozzle
Elevation = 3.96 m
200-micron droplets
Droplet Velocity = 1.4 m/s
Surface Coverage = 14.85 m²/nozzle
0.5 m/s air entrainment rate
Total spray capacity = 19 kg/s

Figure 4.4. Example of a water spray for mitigation of an anhydrous ammonia release.

TABLE 4.3
Maximum Predicted Impact of Water Spray on Dispersion Distances at a
Water to Ammonia Ratio of 5 (water flow is 19 kg/s from all 20 nozzles)

	Unmitigated Scenario	Mitigated Scenario
Distance to 100 ppm (m)	9,000	920
Distance to 1000 ppm (m)	2,000	225
Distance to 10,000 ppm (m)	900	60
K_g, Mass transfer coefficient (kmol/m²/s)	—	0.0192
Ratio of ammonia airborne/ammonia released	1	0.0251
Rate of ammonia airborne (kg/s)	3.8	0.0953

and water contact in a more rigorous manner can be found in Schatz and Koopman (1990), Blewitt et al. (1991), Petersen and Blewitt (1992), Fthenakis and Blewitt (1993) and Fthenakis et al. (1995).

4.5.2. Spray Nozzles

Introduction

The spray nozzles in a liquid curtain must perform two functions. First, they must meter the flow to be sure the liquid reaches all sections of the curtain. Second, at each location they must break up the liquid into droplets.

To ensure that a nozzle meets the above objectives, one must specify the following:

- The flow rate versus pressure characteristics of the nozzle. This information is available in nozzle suppliers' catalogs.
- The type of spray pattern the nozzle is to produce.
- The size of the droplets that are to be produced.
- The size of the nozzle connections to the feed pipe.
- The material that the nozzle is to be made of.

Each of these nozzle parameters will be discussed in the following sections.

Nozzle Flow Rate

The capacity of a nozzle depends on the pressure difference across the nozzle's orifice. This pressure difference is usually a function of the liquid pressure at the nozzle inlet and the atmosphere. However, if the nozzle is to discharge into a volume that is greater than atmospheric pressure, this must be taken into account and the pressure difference adjusted accordingly.

The nozzle capacity data presented in most vendor catalogs are based on the flow of water through the nozzles. The properties of fluids other than water can affect nozzle performance. For instance the flow rate of liquids denser than water will be lower in some spray systems, unless additional energy is supplied to the system to move the denser fluid, that is, a higher horsepower pump.

Viscosity affects nozzle performance, with higher viscosity fluids inhibiting atomization. Generally, fluids with viscosities greater than 100 cps are difficult to atomize unless another fluid like air or steam is provided.

The importance of having a properly designed piping system for feeding liquid to the nozzle is discussed in Section 4.5.3.

Nozzle Spray Patterns

The sprays produced by a nozzle can be categorized by their shapes as conical, fans, or fogs. The spray pattern that will be used is highly dependent on the planned application.

Conical sprays can be further broken down into either full-cone or hollow-cone sprays. In the hollow cone pattern all of the spray is located at the surface of the cone produced by the nozzle and none of it is inside the cone. In the full cone pattern the liquid being sprayed fills the cone pattern produced by the nozzle.

The differences in full- and hollow-cone spray patterns can be seen in Figure 4.5. In this figure the spray flux, gpm/ft^2, is plotted versus the radial distance from the center line of the spray nozzle. For the hollow-cone spray there is very little water at the center of the spray. Instead, it is all at the outer edge. Conversely, for the full-cone spray, the water is fairly evenly distributed radially about the center line of the nozzle.

Nozzles that spray water as a sheet of liquid produce a pattern known as a fan. At the extreme are nozzles that atomize the liquid stream into a mass of very fine droplets. These nozzles are generally referred to as fog nozzles.

HOLLOW CONE SPRAY PATTERN

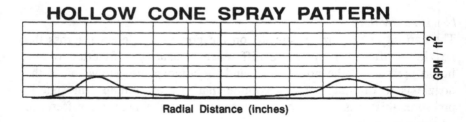

Radial Distance (inches)

FULL CONE SPRAY PATTERN

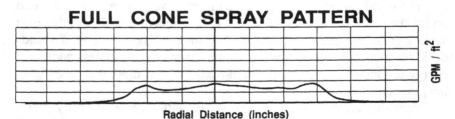

Radial Distance (inches)

Figure 4.5. Spray patterns for hollow- and full-cone spray nozzles. (BETE Fog Nozzle, Inc., 1994).

When nozzles are used in water curtains it is necessary to ensure that the spray pattern produced will be resistant to the prevailing meteorological conditions.

Spray Droplet Size

The size of the droplets produced by a spray nozzle is an essential piece of information as shown in the earlier discussions of water curtains. The droplets must provide adequate surface area for absorption to mitigate a vapor cloud of material having good water solubility. When a water curtain is to be used to mitigate a vapor cloud by diluting it, then the volume of the droplets (i.e., its momentum) is important.

A spray nozzle will actually produce a range of droplet sizes which are expressed as mean or median diameters. The most frequently used mean and median diameters are defined below:

- **Arithmetic Mean Diameter:** The average diameter of all the droplets in the spray sample.

- **Volume Mean Diameter:** The diameter of a droplet whose volume, if multiplied by the total number of droplets, will equal the total volume of the sample.

- **Sauter Mean Diameter:** The diameter of a droplet whose ratio of volume to surface area is equal to that of the complete spray sample.

- **Mass (Volume) Median Diameter:** The diameter that divides the mass (or volume) of the spray into two equal halves. One half of the mass is made up of droplets smaller than this diameter, and the other half of diameters that are larger.

- **Surface Mean Diameter:** A spray droplet whose diameter is equal to the total surface area of all the droplets divided by the number of droplets in the sample.

The Sauter mean diameter has been found to be the most useful of the above definitions for characterizing the spray produced by a nozzle. It is a good indicator of a spray's performance in complex interactions with a droplet's surface and volume. Applications include spray drying, evaporative cooling, dry scrubbing, gas quenching, and gas absorption (Stavis, 1991).

Figure 4.6 shows the range of droplet sizes that can be produced by direct pressure and air atomizing nozzles. In addition it categorizes by size range the different types of particles that may be found in the atmosphere.

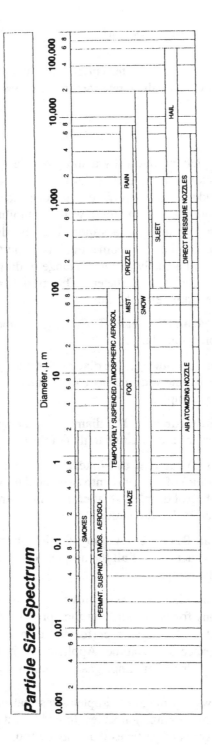

Figure 4.6. Spray nozzle and atmospheric particle size spectrum. (BETE Fog Nozzle, Inc., 1994).

Nozzle Connections

Selection of an appropriate nozzle can be made from a vendor's catalog once the liquid flow (based on the available pressure drop and desired spray pattern and droplet size) has been specified. Once this selection has been made, the size of the connection to the main liquid supply will be known as well.

Material of Construction

Nozzles are available in almost all the common and exotic plastics and metals that are used to fabricate equipment. When selecting the material for a nozzle to be used in a water spray curtain the material must be one that will stand up to the environment in which it will be used. This is critical to be sure that the nozzles will not suffer corrosion to critical dimensions which would result in loss of the needed spray pattern.

4.5.3. Water Supply Capacity, Pressurization, and Reliability

Fundamental to the use of fixed water-spray systems and other water-based extinguishing systems is a water supply that is adequate in terms of capacity, pressurization, and reliability. The capacity available may come from bodies of water, domestic supplies, and/or tank storage. A system's adequacy must be judged on the basis of its projected maximum flows, the duration of its flows, and the volumes necessary to meet industrial needs. The latter is relevant in cases where there is no dedicated supply.

Water pressure is a key consideration in terms of the adequacy of application and system operation and maintenance. While some domestic supplies have "high pressure systems," most are in need of pressurization, as is water from bodies of water and tanks. Fire pumps are routinely used, most often in multiple installations with different means of power drive (e.g., electric, diesel, or steam). When fire pumps operate at high pressures, there is the potential for leaks and line fracture. Periodic maintenance, including hydrostatic testing, will help minimize the potential for such failures.

Most large and complex facilities have multiple sources of water, often from different supplies. Multiple pumps with different power sources are typically installed in parallel to ensure reliable operation in case of a pressure drop. Other reliability considerations include the variability in natural water supplies during different times of the year, for example, fall drought conditions, ice formations, and also the condition of water intakes, for example, trash buildup.

Hydraulic calculations of water supply systems are critical to confirming their adequacy. Often such calculations are not current and, hence, flows and pressures are not readily available to be used as part of evaluations of system performance in cases where different events necessitate their use.

Overall system maintenance is key to ensuring adequate and reliable water flows. Constant water levels, open and operable control valves, and fire pumps that are maintained and tested periodically, are some of the requirements that must be addressed as part of a long-term program.

4.5.4. Fixed Water-Spray Systems

4.5.4.1. Applicability of Fixed Water-Spray Systems

Fixed water-spray systems are a proven means of extinguishing fires and cooling tanks from the effects of thermal radiation. They can also be used to mitigate the effects of flammable and toxic vapor clouds through dilution. This mitigation can be effected by the entrainment of the surrounding air by the spray, absorption or adsorption of the gas by the water drops, or by heat transfer into a cold plume, thereby reducing the vapor cloud's negative buoyancy and assisting in its dispersal (McQuaid, 1977).

The suitability of a water spray for application to releases composed of water-soluble materials is specific to the particular material in question. If applied to cryogenic materials, a water spray can cause the material to heat and violent boiling can result. Therefore, it is prudent to understand the properties of the material in question. Key considerations include the flash point, specific gravity, viscosity, and solubility of the material, the temperature of the water spray, and the temperature of the hazardous material (NFPA 15, 1990).

4.5.4.2. Design of Water-Spray Systems

Since water-spray systems can be applied to different hazardous materials, the designs vary considerably. Critical design features include the water pattern, droplet size, velocity, and density of the spray discharged from the nozzles (NFPA 15, 1990). When a water-spray system is used for vapor mitigation, activation should be automatic. Use for other applications, (e.g., cooling) would require manual activation. A gas-detection system is the usual means of activating a system automatically. A crucial design consideration is the positioning of the detectors and application devices since improper placement can result in downwind vapor travel without system activation.

The water spray system must also be designed to ensure that water does not come in contact with live electrical equipment. Adequate clearances should be maintained under all anticipated weather conditions.

4.5.4.3. Inspection and Preventive Maintenance

As with all water-based extinguishing systems, particularly those installed outside, inspection and preventive maintenance are critical to reliable operation. Detection devices are typically inspected, tested, and maintained quarterly or at a frequency deemed more appropriate for a particular location. The water-spray system itself is typically flow-tested at least annually (NFPA 25, 1995). Because of the nozzles' small orifice size, scale and corrosion buildup can compromise flow and spray patterns. The use of strainers and the conduct of flow tests will help minimize this potential problem. As stated earlier, preventive maintenance of the overall water supply and distribution systems is an integral part of a reliable system.

4.5.5. Monitor Nozzle and Hydrant Protection

Water sprays from monitor nozzles and hose lines can be used for vapor mitigation. Tests have been conducted in which monitor nozzles and hose lines have been used to create a chimney effect through which the gas is forced upward and dispersed at a high elevation (Beresford, 1981). Application techniques and flow rates are facility-, installation-, and material-specific. Careful planning, analyses, and testing should be conducted prior to deciding on the use of a mobile water spray as a proven means of mitigation. Preventive maintenance of this equipment is key to reliable operation. Hose lines, typically, are hydrostatically tested annually. Flow tests should also be conducted periodically.

Figure 4.7 is a photograph of an elevated monitor nozzle located in a refinery for the purpose of mitigating, by water spray, any releases of hydrofluoric acid that might occur within its area of coverage.

4.5.6. Environmental Considerations

4.5.6.1. Climate

As with all water-based extinguishing systems, there are temperature considerations associated with water-spray systems. In cold climates where freezing is a potential problem, selection of a water-spray system should not only be based on potential benefits, but also on the possibility of reliability problems. Care should be taken in the design to ensure that the

Figure 4.7. Elevated water cannon for use in HF release mitigation.

risk of freezing is minimized. In addition, the effectiveness of such systems should be carefully evaluated based on the different ambient conditions that could be experienced.

4.5.6.2 Contaminated Water Run-off

Whenever a water-based system is used for vapor mitigation, there is an obvious concern associated with the water run-off and its potential contamination. Steps must be taken to ensure that there is containment and/or disposal to minimize any adverse environmental effects. Sloping and grading should be such that the run-off is collected in dikes and/or underground drains (which are independent of storm water drains). Care should be taken to ensure that storage capacity is adequate and that the areas are well maintained, for example, no grass and weeds in diked areas. For contained systems with potentially reactive materials (e.g., halogenated acids), one must assure that water entering the contained system is kept to a minimum. Direct water spray into the diked areas should be minimized.

4.6. Vapor-Phase Dilution Systems

4.6.1 Overview

This section describes four additional techniques for mitigating releases of hazardous substances into the atmosphere. In addition to water, both steam and compressed air can be used to promote the movement of air to dilute a hazardous material. Foam can be used as a scrubbing medium to entrap a hazardous material in its structure. Foam scrubbing is often effective for materials that are soluble in water, or that are highly reactive with an additive contained in the foam. Lastly, dry powders can be used to capture a reactive chemical released into the air.

4.6.2. Steam Curtains

4.6.2.1. Application of Steam Curtains
Steam curtains are best used for diluting heavier-than-air releases of flammable vapors, not toxic materials. For flammable materials the level of dilution with air that has to be obtained is the lower explosive limit; toxic materials could require dilution to <100 ppm range. Moreover, while steam curtains can provide the thermal effects that will help disperse flammable material, they hinder the absorption effects needed for toxic materials, especially materials that are water-soluble.

4.6.2.2 Design Considerations
Four design criteria will help to optimize the effectiveness of a steam curtain.

First, a steam curtain should be designed to activate rapidly. As discussed in Chapter 2, a gas cloud develops very quickly after an accidental release, requiring immediate mitigation. With a steam curtain, this can be accomplished by connecting the hydrocarbon or gas detectors to the steam curtain controls and having them activate the system automatically. There should always be a backup system to activate the curtain manually.

Second, a steam curtain should be set up to completely surround the area in which a flammable material release could occur. However, because of wind conditions, it might not be necessary to activate all sections of the curtain, just those downwind of the release. Therefore, it should be possible to control the sections of the curtain that are actually in service through remotely controlled, quick-acting valves. Such control provides the capability to quickly change active curtain sections as wind direction varies. Keep in mind, though, that capability requires good local instrumentation to determine wind direction. Being able to shut down a part of the system that is not needed may be critical to ensuring the continued availability of steam to the curtain.

Third, a steam curtain should be designed to allow quick activation of steam flow to the curtain's piping system. This means that the steam lines from the headers to the steam curtain sections must be well trapped to ensure that when the valves open, hot dry steam is supplied to the header. A delay while the condensate is drained from the system, or the unstable operation of the curtain that could result, is not desirable. Also, having an adequate supply of steam available for the time required to bring other mitigation techniques into play to secure the situation is essential.

Fourth, the piping used to provide the actual curtain should be placed on a low wall of light construction about 1.5 m high. The wall is, in effect, a vapor fence that slows the spread of a vapor cloud and enhances the vapor detection of the monitors. Also, as described in the following section, it provides some mitigation until the steam curtain can be brought into service. Lastly, having the piping on a wall at this height keeps out dirt.

One steam curtain described in the literature (*Guidelines for Vapor Release Mitigation,* CCPS, 1988), is a 6-in. diameter pipe with a row of $5/32$-in. holes spaced at 4-in. intervals. Steam was supplied to this curtain at 250 psig. It was effective in reducing concentrations by a factor of 30 when the steam flow rate was equal to the vapor release rate.

Steam curtains consume steam at a high rate. One reference (*Guidelines for Vapor Release Mitigation,* CCPS, 1988) quotes 0.2 lb of steam per pound of material released, another quotes 100 lb of steam at 250 psig per hour per foot of curtain. Because of these high levels of steam consumption, steam curtains are usually confined to protecting a small area or places where small leaks might occur.

An example of the use of steam curtains around a small area may be found in the hydrotreating processes used in oil refineries. Pipes or tubing rings with holes positioned on the inside of the ring are placed around large flanged joints in the parts of the system that handle hydrogen. Steam is then piped to the rings to dilute any small quantities of hydrogen that may leak from the flange joints, thus ensuring dispersion.

Finally, although the generation of static electricity in steam curtains has been a concern, investigations have indicated that this is not a problem as long as the system is well grounded (Lees, 1980).

Since a steam curtain may activate at any time, warning alarms and signs should be placed to alert personnel who may be in the area to the need for immediate evacuation. Also, personnel who may have to work close to the curtain in an emergency should be provided with protective equipment to prevent being scalded by the hot steam or condensate droplets with which they may come in contact. Naturally, nonessential personnel should be kept out of areas affected by steam curtain operation.

4.6.2.3. Inspection and Preventive Maintenance

When a steam curtain is to be used to mitigate the consequences of a flammable material release, it must be treated as if it were part of the site's firefighting system. This means that a managerial system must be in place to ensure that the steam curtain system is inspected periodically. Deficiencies should be reported and repaired promptly. Follow-up investigations must be made to ensure that the inspections, deficiency reporting, and repairs occur in a timely manner. If the system must be shut down for the repairs, the same procedures that would be followed for a sprinkler or deluge system shutdown should be utilized.

Items that should be periodically inspected are:

- Gas sensors to ensure they are functioning and correctly calibrated;
- Entire control system, to assure it functions properly;
- Curtain piping, to ensure that all holes are clear;
- Manual valves in the steam lines, to ensure that the curtain steam supply is open and that proper management controls, such as locking, tagging, or car sealing, are being followed; and
- All supply line steam traps, to ensure that they are functioning.

Many sites operate steam trap inspection programs as part of their energy conservation effort. Even if the traps for the steam curtain are included in this program, they should also be checked as part of the curtain inspection program.

The hydrocarbon or gas detectors used to activate the curtain system should also be calibrated and tested periodically, in accordance with the supplier's recommendations.

In addition to the inspections cited above, the steam curtain should be tested once a year. Operating the system will help to ensure that no faults were overlooked in the more frequent system inspections.

4.6.3. Air Curtains

An air curtain works somewhat like a water or steam curtain, but offers no potential to absorb a toxic material or increase dispersion from thermal effects. An air curtain merely provides air movement that promotes air dilution.

There has been little experience with the testing and use of air curtains so data on their viability as a mitigating measure (Rulkens et al., 1983) are limited. In considering design of an air curtain, the availability and reliability of a significant supply of air are of critical importance, and may be the limiting factor.

4.6.4. Foam Scrubbing

A foam application, still in development, uses a scrubbing agent to remove airborne releases of toxic substances or particulate matter. This technique utilizes the air containing the toxic or particulate matter to generate foam. The foam then encapsulates the hazardous material in bubbles with a large interior surface area, where absorption into the liquid shell of the foam bubble can occur. The agent in the foam can then react with hazardous gases, neutralizing and rendering them harmless. As the foam collapses, it can be drained from the area in which it was formed and then processed for environmentally safe disposal. This approach is described in Brown et al. (1990) which summarizes past work and recent research on this technique.

The foam scrubbing technique is effective because it brings the hazardous material into close contact with the foam by getting it into the bubbles. This is different from using a foam blanket as a cover for spills (see Chapter 3). With the large internal surface area of the foam available for absorption or mass transfer, an equilibrium concentration between the contaminated air inside the bubble and the foam cell wall liquid can be developed rapidly. Unabsorbed gas that is still in the foam bubbles when they collapse is released. This results in the slower release of a smaller quantity of hazardous material, which should result in a reduced hazard zone downfield.

Data from Brown et al. (1990) show that if the foam contains reactive materials, these materials react to form a neutralized compound as they are absorbed into the liquid bubble. The data indicate that in this situation, the foam may completely recover the released material if adequate amounts of the reactant are contained in the foam when it is generated. The reported work also indicates that heat effects from neutralization reactions do not weaken the stability of the foam.

The foam technique requires two main components: a foam generator, and a pressure source which feeds the contaminated air or vapors to the foam generator. Three applications of this technique for controlling releases are described in Brown et al. (1990). The first works with releases that occur inside a process building. Here, the ventilation system brings the contaminated air to the foam generator, which entraps the hazardous material in the foam. A second application places a foam generator in the tail pipe of a pressure safety valve (PSV) that discharges to the atmosphere. When the valve lifts, the foam generator activates and the foam captures and scrubs the discharged material. The third technique uses a mobile system that captures the hazardous release from a leak or other source with a local ventilation system and then discharges it through a foam generator.

The foam-scrubbing technique has been studied for a variety of gases, including ammonia, sulfur dioxide, hydrogen sulfide, formaldehyde, and nitrogen oxides. Liquid and solid streams that have been experimented with include sulfuric acid fog and mist, dust, fly ash, and methylene blue.

4.6.5. Dry Powder Curtains

In cold climates, especially when temperatures are below freezing, a water curtain may suffer operating problems when ice forms at the spray nozzles or the water drops produce snow. Either of these occurrences could result in a significant reduction in the effectiveness of a water curtain.

It is known that dry powders, such as sodium bicarbonate, calcium chloride, magnesium oxide, and calcium hydroxide, readily react with hydrogen fluoride. Use of similar materials for the control of liquid spills was discussed in Chapter 3 of this guideline.

A recent small-scale study (Schatz, 1993) evaluated several powders—either metal carbonates, oxides, or hydroxides—to determine reaction rate constants and applicability for mitigating an HF cloud. Several powders were found to have sufficiently high reaction rates and HF removal efficiencies to make them feasible for HF mitigation. Powder weight ratios as low as twice the stoichiometric ratios achieved the same HF removal efficiency (90%) as the highest water-to-HF ratios (40:1) used in the spray tests described in Section 4.3. Upon reaction with HF, the dry powders usually form a wet, nontoxic mud that is easy to collect and remove.

The technology for applying dry powders for firefighting purposes is well developed for small-scale situations, such as restaurant kitchen firefighting systems. (A scale-up to large industrial systems has not been done as yet, but no inherent technical problems would preclude such a scale-up.) However, to prevent lump formation, powder must be carefully stored and kept dry. The system should also receive a complete performance test periodically, to ensure that it will function when needed without having to discharge the powder inventory.

Using dry powders for the mitigation of airborne or liquid HF and other toxic materials is possible. The technique would work well where the maximum spill is known in advance so that stationary dry powder systems of finite, yet sufficient, capacity could be designed.

4.7. References

Beresford, T. C., 1981. The Use of Water Spray Monitors and Fan Sprays for Dispersing Gas Leakages. *Northwest Branch Papers*, 5: 6.1–6.3. London: Institution of Chemical Engineers.

BETE Fog Nozzle, Inc., Manual No. 104, Greenfield, MA: BETE Fog Nozzle, Inc.

Blewitt, D. N., J. F. Yohn, R. P. Koopman, T. C. Brown, and W. J. Hague, 1987. Effectiveness of Water Sprays on Mitigating Anhydrous Hydrofluoric Acid Releases. *Proceedings of the International Conference on Vapor Cloud Modeling*, pp. 155–171. New York: American Institute of Chemical Engineers.

Blewitt, D. N., R. L. Petersen, M. A. Ratcliff and G. Heskestad, 1991. Evaluation of Water Spray Mitigation System for an Industrial Facility. *Proceedings of the International Conference and Workshop on Modeling and Mitigating the Consequences of Accidental Releases of Hazardous Materials*, pp. 483–510. New York: American Institute of Chemical Engineers.

Briffa, F. E. J. and N. Dombrowski, 1966. Entrainment of Air into a Liquid Spray. *Journal of American Institute of Chemical Engineers*, 12(4): 708–717.

Brown, P. M., R. H. Hiltz and J. E. Brugger, 1990. Detoxification of Released Vapors/Particles by Entrapment in Chemically Active Foam. *Proceedings SuperFund 1990*, pp. 589–594. Washington, D.C.

Carroll, J. J, 1991. What Is Henry's Law? *Chemical Engineering Progress*, September 9: 48–52.

CCPS (Center for Chemical Process Safety), 1988. *Guidelines for Vapor Release Mitigation*. New York: American Institute of Chemical Engineers.

Donaldson, C. and C. Snedickar, 1971. A Study of Free Jet Impingement. *Journal of Fluid Mechanics*, 45: 281–319.

Eggleston, L. A., W. R. Herrera, and M. P. Pish, 1976. Water Spray to Reduce Vapor Cloud Spray. *Journal of Loss Prevention*, 10: 31–42.

Emblem, K. and O. K. Madsen, 1986. *Full Scale Test of a Water Curtain in Operation*.

Fthenakis, V. M., 1989. The Feasibility of Controlling Unconfined Released of Toxic Gases by Liquid Spraying. *Chemical Engineering Comm.*, 83: 173–189.

Fthenakis, V. M. and V. Zakkay, 1990. A Theoretical Study of Absorption of Toxic Gases by Spraying. *Journal of Loss Prevention in the Process Industries*, 3: 197–206.

Fthenakis, V. M. and K. W. Schatz, 1991. Numerical Simulations of Turbulent Flow Fields Caused by Spraying of Water on Large Releases of Hydrogen Fluoride. In J.W. Hoyt and T.J. O'Hern, Eds., *Fluid Dynamics of Sprays*, FED-131. Pp. 37–44. New York: ASME (American Society of Mechanical Engineers).

Fthenakis, V. M., K. W. Schatz, and V. Zakkay, 1991. Modeling of Water Spraying of Field Releases of Hydrogen Fluoride. *Proceedings of the International Conference and Workshop on Modeling and Mitigating the Consequences of Accidental Releases of*

Hazardous Materials, pp. 403–427. New York: American Institute of Chemical Engineers.

Fthenakis, V. M. and D. N. Blewitt, 1993. Mitigation of Hydrofluoric Acid Releases: Simulation of the Performance of Water Spraying Systems. *Journal of Loss Prevention in the Process Industries,* 6(4): 209–218.

Fthenakis, V. M. and D. N. Blewitt and W. J. Hague, 1995. Modeling Absorption and Dilution of Unconfined Releases. *Proceedings of the International Conference and Workshop on Modeling and Mitigating the Consequences of Accidental Releases of Hazardous Materials,* pp. 573–592. New York: American Institute of Chemical Engineers.

Gutmark, E., M. Wolfshtein and I. Wygnanski, 1978. The Plane Turbulent Impinging Jet. *Journal of Fluid Mechanics,* 88(4): 737–75.

Heskestad, G., H. C. Kung, and K. F. Todtenkopf, 1976. Air Entrainment into Water Sprays and Spray Curtains. *ASME Winter Meeting,* 76: 1–14. New York: American Society of Mechanical Engineers.

Lees, F. P., 1980. *Loss Prevention in the Process Industries.* 2 vols. London and Boston: Butterworths.

Looney, M. K. and J. J. Walsh, 1984. Mean-flow and Turbulent Characteristics of Free and Impinging Jet Flows. *Journal of Fluid Mechanics,* 147: 397–429.

Martinsen, W. E., S. P. Muhlenkamp, and L. J. Olson, 1977. Dispersing LNG Vapors with Water. *Hydrocarbon Processing* (July): 261–266.

McQuaid, J. and R. D. Fitzpatrick, 1981. *The Uses and Limitations of Water Spray Barriers.* London: Health and Safety Executive

McQuaid, J., 1977. The Design of Water-Spray Barriers for Chemical Plants. *Proceedings of the International Symposium on Loss Prevention and Safety Promotion in the Process Industries.* September 6–9, Heidelberg, Germany.

Meroney, R. N. and D. E. Neff, 1984. Numerical Modelling of Water Spray Barriers for Dispersing Dense Gases. *Journal of Boundary-Layer Meteorology,* 31 (March): 233–247.

Moodie, K., 1985. The Use of Water Spray Barriers to Disperse Spills of Heavy Gases. *Plant Operations Progress,* 4 (October): 234–241.

Moodie, K., 1981. Experimental Assessment of Full-Scale Water Spray Barriers for Dispersing Dense Gases. *Northwest Branch Papers,* 5. London: Institution of Chemical Engineers.

Moore, P. A. C. and W. D. Rees, 1981. Forced Dispersion of Gases by Water and Steam. *Northwest Branch Papers,* 5: 4.1–4.14. Institution of Chemical Engineers.

NFPA 15, 1990. Standard for Water Spray Fixed Systems for Fire Protection, Chapter 1. Quincy, Massachusetts: National Fire Protection Association.

NPFPA 25, 1995. Inspection, Testing, and Maintenance of Water-based Fire Protection Systems, Chapter 7. Quincy, Massachusetts: National Fire Protection Assocation.

Petersen, R. L., and D. N. Blewitt, 1992. Evaluation of Water Spray/Fire Monitor Mitigation Systems for Two Refineries. *Proceedings of the 1992 Process Plant Safety Symposium.* Vol 1, pp. 477–500. New York: American Institute of Chemical Engineers.

Prugh, R. W., 1986. Mitigation of Vapor Cloud Hazards, Part II: Limiting the Quantity Released and Countermeasures for Releases. *Plant Operations Progress,* 5(3, July): 169–174.

Rulkens, P. F. M., A. P. Burger, G. P. TenBrink, and C. P. Gullemond. 1983. The Application of Gas Curtains for Diluting Flammable Gas Clouds to Prevent Their Ignition. In *4th*

International Symposium on Loss Prevention at Harrogate. 80: F15–F25. London: Institution of Chemical Engineers Symposium

Schatz, K. W., 1993. Personal communication to Arthur D. Little, Inc.

Schatz, K. W. and R. P. Koopman, 1990. Water Spray Mitigation of Hydrofluoric Acid Releases. *Journal of Loss Prevention in the Process Industries,* 3: 222–233.

Stavis, J. P., 1991. "Practical Considerations for Nozzle Selection in Water Spray HF Mitigation Systems." Paper presented at Petro-Safe '91 Conference, Houston, February.

Van Doorn, M., 1981. "The Control and Dispersion of Hazardous Gas Clouds with Water Sprays." Ph.D. diss., Department of Applied Physics, Delft University of Technology, Netherlands. NTIS DE82-902736.

Van Zele, R. L. and R. Diener, 1990. On the Road to HF Mitigation. *Hydrocarbon Processing.* (June): 92–98.

Watts, J. W., 1976. Effects of Water Spray on Unconfined Flammable Gas. *Journal of Loss Prevention,* 10: 256.

5

Secondary Containment

5.1. Introduction

Containment uses a physical barrier to prevent an uncontrolled release of materials to the environment. The walls of a vessel or pipe serve as the primary containment barrier that encloses harmful materials. Redundant (secondary) containment serves as a safeguard if the primary barrier fails, and is considered a postrelease mitigation measure. Containment can take many forms, depending on factors such as the system or process to be contained, the risks involved with a release, and the cost benefit of the additional secondary containment.

5.2. Diking

The primary objective in placing dikes around vessels that contain flammable and/or toxic materials is to alleviate the potential hazards from an accidental release of the fluid by reducing the total vaporization rate to the atmosphere through a reduced surface area. The design of a dike should consider the impact on incident mitigation. A spilled fluid can be safely contained by conducting a review of dike design options so that hazards to equipment, personnel, and the surrounding communities can be minimized.

Diking is a common industry practice for spill retention around flammable liquid storage vessels. Regulations and codes require that exclusion zones be placed between storage vessels to maintain the structural integrity of nearby vessels or process equipment in the event of a fire within the dike (Welker, 1987). Studies of flammable vapor boiloff rates as a function of dike design have been used in the natural gas industry to evaluate optimal materials of construction (Arthur D. Little, Inc., 1974). In recent years, there has been an increased emphasis on the secondary containment of toxic liquids as well. The focus has been on dike design methods that can reduce the toxic vapor generation rate and/or its duration, thereby mitigating the attendant toxic hazards following a release (Dilwali and Mudan, 1987; Harris et al., 1987).

Dike design alternatives include variations in dike geometry, consideration of materials of construction, and provisions for removing material from a dike. Each of these issues is discussed in the subsections below. Specific regulatory code requirements, as they apply to secondary containment, are also introduced.

5.2.1. Optimal Dike Geometry

Geometric considerations in dike design include the base area configuration, slope of the dike floor, and the height of a retaining wall. Optimal dike design must satisfy two criteria: (1) it must provide safe fluid containment, and (2) it must be proportioned for the most economical size.

The configuration of a level impoundment base area is generally square, rectangular, or circular. Specific empirical solutions for these three geometries have been developed which minimize the costs associated with a dike of known volume requirements (Mazzarella, 1984; Welker and Johnson, 1987).

By minimizing the surface-area-to-impoundment-volume ratio, the total base plus wall area available for boiloff can be reduced. This reduction, in turn, leads to lower vapor generation rates and downwind hazards. It should be noted, however, that these designs are typically driven by cost optimization rather than safety.

Optimally, sizing dikes from a safety perspective also requires a thorough assessment of applicable regulations, available land area, topography, stipulated volume, and dimension constraints. Conventional computations involve trial-and-error procedures, which can be computerized, to estimate dike height, cross-sectional base area, and base width required to retain a specified liquid volume (Hangs, 1986). For circular base geometries, nomographs can also be used to estimate the safety of diked-in areas quickly (Zanker, 1979). The geometry that results in the lowest boiloff rate and/or duration is the optimal dike configuration.

Many diked areas are unnecessarily flat and level, which allows a liquid spill to spread over the entire surface (CCPS, 1988a). The optimal design for a sloping dike is a simple, one-directional slope away from the tank, leading to a trench or sump that allows for the possibility of remote collection of spills (Arthur D. Little, Inc., 1974). The ultimate effect of a sloped dike is to reduce a flammable liquid inventory available for a fire, or to reduce the available spill surface area (i.e., vapor generation rate) for a toxic liquid spill.

Dike wall heights (e.g., curbs) can range from a few inches to several feet. High-walled impoundments may be justifiable where limited space is available, and where the impacts of a vapor release may be severe (Harris et al., 1987). Since the surface-area-to-volume ratio for vaporization is lower for the same volume, a high-wall dike will generally be more effective at reducing vapor evolution rates. However, it is worth noting that for cryogenic liquid containment, an increase in dike wall height can result in increasing the surface area of the dike itself and area available for heat transfer (Dilwali and Mudan, 1987). Furthermore, use of high-walled impoundments can hinder the ability to routinely inspect the exterior of the tank wall. The regulatory requirements concerning dike design are detailed in Section 5.2.4.

Instead of high retaining walls, vapor fences can be added to the top of a dike as an effective means of providing increased containment and a higher point of vapor release (Harris et al., 1987). Another approach to mitigating downwind dispersion hazards involves the use of "leaky" high-dike vapor fences placed on top of a normal dike. This design feature allows a slow, controlled vapor bleedoff in the event of a spill. For the same mass involved, a continuous release leads to significantly smaller downwind hazards than a larger, short-duration puff (Arthur D. Little, Inc., 1974).

5.2.2. Materials for Dike Construction

Since the purpose of diking is to contain a spilled fluid effectively while reducing the vapor generation rate, a secondary containment system that will be impervious to the fluid must be provided. It should also have favorable thermal properties to retard the boiloff rate.

Dikes may be made of earth, steel, crushed stone, or concrete, and they may be lined with a layer of clay or asphalt, or plastic film. There are several constraints on dike material; it must be chemically compatible with the process material, be durable under normal weathering conditions, have structural characteristics suitable to support loads from maintenance vehicles, and be cost effective. Ideally, to prevent ground contamination, the ground between a vessel and a dike wall should be nonporous.

The thermal properties of a dike can have a profound effect on the vapor boiloff rate, especially for materials with boiling points below ambient temperature (i.e., refrigerated spills). For a given base area, the vapor generation rate is a function of the rate of heat transfer to the liquid pool. The total heat transfer rate has three components: (1) conduction effects and the resultant cooling of the contact surfaces; (2) convective heat from the wind; and (3) solar radiation. During a release it is the conductive heat transfer rate (i.e., the rate of thermal energy transferred from the dike to the spilled liquid), along with the sensible heat buildup in the spill surface and substrate, that is important. Therefore, it is important to consider the heat balance both before and during the spill in selecting dike-insulating materials. The effects of refrigerated spills on dike surfaces has been studied for flammable materials (Drake and Reid, 1975; Moorhouse and Carpenter, 1986), as well as toxic materials (Dilwali and Mudan, 1987; Feind, 1974).

Vapor generation rate is proportional to the heat transfer parameter, which is a function of the dike material properties. Table 5.1 (Dilwali and Mudan, 1987) lists values of the heat transfer parameter for various dike materials. It is evident that insulating the dike floor and walls significantly reduces the heat transfer rate and the resultant vapor evolution rate. Significant advances have been made in the development of low-cost polymer concrete for dike insulation (GRI, 1990). In addition to its low thermal diffusivity, insulating polymer concrete exhibits good load-bearing strength properties, impermeability to water, and long-term stability under extreme environmental conditions. However, before any of these insulating materials are used, their chemical compatibility with the materials they may come in contact with should be checked.

For chemicals with boiling points above ambient temperatures contained in a dike, the wind speed across the surface of the material becomes important.

5.2.3. Provisions for Removal of Materials from a Dike

Diking provides an initial means of safely containing a hazardous material spill. Unless and until the spill contents are removed from the impoundment enclosure or some other action is taken, release to the atmosphere will continue, and the potential for toxic exposure, pool fire, and/or explosion will remain. The addition of a collection pit or drainage system, in conjunction with a sloping spill-retention area, is an effective means of further reducing the postrelease hazard, as it prevents large pools from forming and thereby reduces the atmospheric exposure of the spilled material. In such a system, a sloping dike routes the spilled fluid, via channels or drains, to a covered basin or underground tank, from which the liquid may either be recovered or treated by pumping, chemical reaction, or absorption.

It is common practice to equip a drainage system or sump with a pump to return the collected inventory to storage or process facilities. Alternatively, for certain toxic materials, the spill is sometimes altered chemically to a nonhazardous substance by draining the spillage to a basin filled with a neutralizing slurry where the reaction forms an insoluble sludge. For acids, in particular, soda ash, limestone, or weak caustic solutions may be used. The reacted product becomes a solid that is neutral and can be disposed of accordingly.

Another technique for removing spillage from a diked area is the use of an all-purpose sorbent. The effectiveness of using sorption as a removal technique is influenced by factors such as the physical and chemical attraction between the spilled material and the sorbent, the surface geometry and area of the sorbent, the contact time between the materials, and the density ratio of the spilled fluid and the sorbent (Bauer et al., 1975). Some effective sorbents include polymethacrylate foam resins, propylene fibers, molecular sieves, expanded clays, polyolefins, polymethylmethacrylates, and polystyrene sulfonates (Bauer et al., 1975).

TABLE 5.1

Values of K for Various Dike Materials (Dilwali and Mudan, 1987)

Material	Bulk Density (kg/m^3)	Heat Transfer Parameter* $[k/\rho C_p]^{1/2}$
1. Soil (dry)	1300–1800	2570
2. Sand (dry)	1600	2660
3. Sand (wet; 3% moisture)	1500	2335
4. Uninsulated concrete	3100	3750
5. Insulated concrete •Dycon K-23™ •Dycon K-3™ •Grace G-24™ •Grace G-34™	290–370 510 510 545	238 335 230 440
6. Polyurethane	120	140
7. Other insulating materials •Celoam™ •Foamglas™	200 140	99 74

*$[k/\rho C_p]^{1/2}$ is a thermal diffusivity parameter, where k = thermal conductivity, ρ = bulk density, and C_p = specific heat.

More economical options include straw, seaweed, leaves, and corncobs. The most versatile material for sorption is activated carbon.

Recent development of sorbents include nontoxic and nonflammable materials that can absorb acids, alkalis, hydrocarbons, pesticides, and chlorinated compounds (*Plastics World,* 1987). These nonreactive sorbents can be placed directly on or around a spill to soak up the liquid. The used sorbent tubes or "pillows" can then be stored temporarily in heavy-duty spill disposal bags for subsequent treatment and disposal.

5.2.4. Regulatory Requirements Regarding Diking

There are several specific regulatory documents governing the siting, design, and capacity of impoundment systems. For flammable materials, these requirements are outlined in the LNG: Federal Safety Standards, 49 C.F.R. § 193 (1990) and the National Fire Protection Association's Flammable and Combustible Liquids Code, NFPA 30 (1990). For toxic materials, similar design guidelines are specified in the sections of the Code of Federal Regulations (C.F.R., EPA Section 32 and EPA Section 40).

The Federal safety standards included in 49 C.F.R. § 193 (1990) define four classes of impounding systems ranging from dikes constructed within 24 inches of the component served to remote impounding spaces (see 49 CFR § 193.2153). The structural requirements specify performance reliability and integrity as a result of imposed loading caused by a full liquid head of spilled material, erosive spill action, thermal gradients, fire exposure, and catastrophic rupture of storage or transport vessels into or near the system (see 49 C.F.R.§ 193.2155).

The NFPA 30 (1990) code specifies design and capacity criteria for both remote impoundments and dikes around tanks (see Section 2.6.1). Remote areas require sloping dikes and safe routing of drainage systems. For tanks within an impoundment, the volumetric capacity must exceed the greatest amount of liquid that could be held by the largest tank, allowing for volume occupied by other tanks within the diked area. Dike average heights may exceed 1.83 m (6 ft), with provisions for normal and emergency access to tanks and equipment. The minimum distance required between the tanks and the toe of the interior dike wall is 1.52 m (5 ft). Several stipulations regarding subdivisions of dikes for multiple tank storage are also detailed in these codes.

Part of the Environmental Protection and Enhancement Act, 32 C.F.R. § 650 (1992), specifies minimum plan requirements for the spill control of oil and hazardous substances. Preventive containment measures are recommended, such as impervious dikes and berms, curbs, spill diversion and retention ponds, and gutters or other drainage systems. The regulations also allow the use of sorbents for containment if structural installation is not practicable. The Environmental Protection Agency has established general design requirements for surface impoundments (40 C.F.R. §267 [1990]). For example, each impoundment must be designed to prevent overtopping from either wind or wave action, overfilling, precipitation, and such. Moreover, structural integrity of the dike must not be compromised from the effects of plants or burrowing animals on earthen dikes, nor from water or wind erosion. Finally, dikes must be installed with impervious liners so that there is no leakage of the hazardous material into the underlying or surrounding soils.

5.2.5. Emergency Response Dikes

In the event of an unconfined spill from a leaking tank truck or tank car, for example, requiring short-term containment until recovery or disposal methods are in place, emergency impoundments can be constructed of earthen material, crushed stone, or gravel. Two important considerations regarding temporary diking include material compatibility and ignition hazards. For reactive materials, it is necessary to establish the chemical/physical compatibility between the dike and the fluid spilled. For example, if materials that react with water are contained in an earthen dike that has a high moisture content, then toxic fumes may be generated, which may prove hazardous to construction personnel. The other possibility to consider is the presence of potential ignition sources during dike construction. For example, heavy construction equipment, such as bulldozers, operated around a flammable spill could ignite the vapors downwind of a release if the concentration falls within the flammable limits.

To ensure timely action in the construction of an emergency dike, an attempt should be made to identify these possible situations and preplan for their occurrence. This preplanning should include a source of suitable machinery and diking materials.

5.3. Double-Wall Containment

For chemicals that pose severe release hazards, one effective but costly way to provide spill containment is to use a double wall on the vessel or piping (CCPS, 1988b). Often construction materials for the outer (secondary) wall are the same as those of the primary inner wall. In case of inner-wall failure, the outer wall will have to withstand the temperature, pressure, and corrosivity of the process fluid. Lower design conditions for the outer wall may be acceptable, depending on the type of annular space monitoring provided for detecting primary failure, and the length of time the outer wall is expected to function without breach.

Without a method or mechanism to detect a catastrophic, or even a small release from an inner wall, the use of a secondary outer wall is superfluous. A failure in the inner wall will certainly lead to a failure of the outer wall if the failure goes unnoticed and no action is taken. The secondary wall may delay the release, but release would be inevitable without annular space detection. Typical detection mechanisms include gas analyzers or pressure detectors for vapors, conductivity switches for liquids, or "weep" holes routed to drain systems that are periodically monitored (Prugh, 1992). A purge gas is sometimes used as a detection medium, to reveal a change in contaminants in the purge gas exhaust (CCPS, 1988b).

Figure 5.1 is a simplified schematic drawing of a double-wall pipe. The hazardous material flows in the inner pipe. The annular space is purged with a flow of nitrogen gas. The discharge of the nitrogen from the annular space is analyzed to detect a leak of the inner pipe. Note that the inlet and outlet valves for nitrogen to the annular space are car-sealed-open (CSO) or locked open (LO).

If a leak does occur in the inner wall, a mechanism must exist to vent or drain the substance between the walls to a safe location. Connections must be provided to neutralize, clean, and hydrotest the annular space after a failure. The annular space must also receive preventive maintenance to assure its proper operation. Periodic checks of the integrity and reliability of leak detection systems should be required.

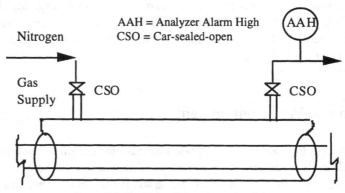

Figure 5.1. Double-wall containment—piping.

The design of double-walled tanks requires some special considerations. For instance, some large cryogenic ammonia tanks are enclosed in secondary containers with leak-detection devices placed in the annular space (CCPS, 1988a). The Chemical Industry Association (of the United Kingdom) recommends a variation in double-containment design where only refrigerated vapors are present in the annular space between the two walls (Chemical Industry Association, 1975).

American Petroleum Institute (API) Standard 620 (1985) also discusses double-wall tanks containing low-pressure liquefied hydrocarbon gases and refrigerated products. Designs for these tanks consist of an inner tank for storing the refrigerated liquid and an outer tank for enclosing an insulation space around the inner tank. The double-wall tank is a composite tank; its inner and outer walls are not necessarily identical in design.

Indeed, the outer wall may need to be stronger to withstand hydrodynamic loads following a sudden failure of the inner wall.

Figure 5.2 is an illustration of a vessel double-wall containment system. The annular space is kept at a higher pressure than the tank. Should a leak develop, the pressure in the annular space will fall because the flow capacity of the nitrogen pressure control system has been limited. A low pressure will occur, tripping a pressure alarm low (PAL), and alerting the operator to the leak.

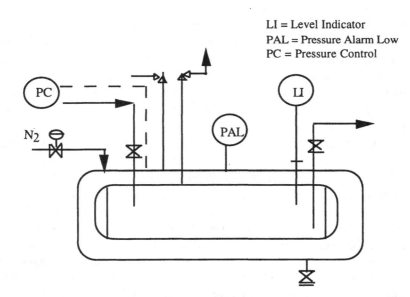

Figure 5.2. Double-wall containment—vessels.

5.4. Enclosures

Enclosures are another form of double containment that prevent (or delay) a release from reaching the environment until proper action can be taken to stop, and drain or vent, the material in a controlled manner (CCPS, 1988a). Use of enclosures is common where other containment systems would be impractical, as in the case of vapor or volatile liquids. In areas of severely cold weather, where most of a process may already be indoors, a form of enclosure already exists to some extent and certain containment principles already apply.

Enclosures are typically designed for low pressures in the inches-of-water range. Other designs, such as those for nuclear reactor containment, are rated much higher, but cost much more. To function properly as an enclosure, a structure should be airtight and able to be purged and vented if a release occurs.

Releases from enclosures are either scrubbed before being released to the atmosphere, vented to a safe location, or routed to a flare system. In all cases, when determining the size and type of vent, maximum release rates and back-pressures while venting should be calculated. An airtight enclosure could structurally fail because of a pressure buildup from liquid vaporization if it is not properly vented during a release (Harris, 1991).

Figure 5.3 is a photograph of a building that contains a process using toxic materials. As can be seen, the building totally encloses the process. Ducts from each of the floors are connected to the scrubbers. This ductwork can be seen on the front of the building and in the lower left side of the photograph (entering the scrubbers). Two scrubbers are provided for this facility. The smallest, on the right, is used to remove any fugitive emissions that might come from the process during normal operation. Should a major release occur in the process building, the larger scrubber, on the left, would then come into service.

An enclosure may also create a safety hazard because it is a confined space. Of concern are asphyxiating gases, such as nitrogen purges, in improperly vented enclosures. If personnel are allowed to enter enclosures, then oxygen analyzers should be used and safe entry practices should be followed. Air connections, for air masks with escape bottles (29 C.F.R. §1910.146), should be located inside each entry door to provide breathing air for the operator and to ensure that negative pressure in the enclosure is maintained while personnel are working inside it. Equipment layout within the enclosure should be kept simple to ensure that air hoses will not become tangled. This is important for quick egress from the enclosure, as are escape bottles, for cases where the breathing air system could fail.

Figure 5.3. Totally enclosed process showing ductwork and scrubbers.

In an enclosure that operates under slightly negative pressure, or in one that goes to negative pressure when a gas is detected and fans or blowers start, at least one of the escape doors should open inward. Negative pressure inside the enclosure will make opening the door more difficult in an emergency. A knock-in panel in the door will help circumvent this drawback.

As with double-wall containment, enclosures are most effective when used with a leak-detection system. These systems should be continuously monitored and connected to both local and control room alarms as the system dictates (CCPS, 1988a). The monitoring and alarm instrumentation should be tested periodically to ensure that it will function as designed. This is discussed further in Chapter 6.

Before designing enclosures for processes involving flammables, the consequences of poor ventilation that could concentrate the release above the lower flammable limit should be considered. Even if there are no ignition sources, continuously monitored combustible gas detectors should be placed in the enclosure. If all other sources of containment for flammables are inappropriate and enclosures are the only option, a means for extraordinary ventilation of even the smallest release should be provided. In addition, NFPA 68 (1994), *Guide for Venting of Deflagrations*, should be consulted and incorporated into the design of the enclosure. This will mitigate the consequences from ignition of the flammable should the other measures mentioned earlier fail.

Enclosures should also be designed with features for collecting liquids from releases, and for equipment drainage or area washdowns. Drains could be routed to enclosed sewer systems or adjacent sumps. Releases should be removed from the enclosure as quickly as possible to minimize vapor buildup or the risk of fire (Harris, 1991).

5.5. Transfer Vessels

A transfer vessel is a device that receives the contents of another vessel for emergency or nonemergency purposes. It can be as simple as a vacuum truck or as complex as a hard-piped, dedicated system. For liquids, the system typically consists of a container or containment system located below the protected vessel where gravity will promote a rapid transfer. In the few instances where a transfer vessel is used with gases, it assists in the depressurization of a process. In other instances, it may consist of a spare vessel capable of accepting the contents of a nearby vessel (in case of fire or leak) so that the damaged vessel's entire contents are not destroyed or released (Lees, 1980). In this case, a pump may be used to make the transfer between vessels.

The storage capacity of a transfer vessel should equal or exceed that of the vessels it is protecting. One approach would be to design a tank equal in volume to the largest storage tank. Because it is often uneconomical to maintain a large, unused storage tank, another approach would be to analyze the risks based on tank levels and design the transfer tank based on average or expected volumes. For a reactor, the transfer vessel size should exceed the reactor's volume, to allow for heat and pressure effects and for any residual feed stream volume.

As with double-wall containment systems, a transfer vessel's construction materials, design pressure, and temperature rating should at least equal those of the equipment being protected. Construction materials can differ if the transfer vessel will only be exposed to the corrosive process for an acceptably short duration.

The design of the transfer system depends on the required flow rate of liquids and vapors into the transfer vessel. If a runaway reaction has to be rapidly dumped to prevent equipment damage or a catastrophic incident, then complete transfer should be effected within a matter of seconds. If the transfer vessel is designed to assist in the reduction of inventory in case of a leak, the transfer may take several hours, during which time the leak can be temporarily patched or the release contained or minimized.

The preferred transfer mechanism uses the potential energy of pressure or gravity. If the transfer operation is critical, then system reliability is highly important, and the transfer mechanism should be fail-safe. Where time allows, pumps, compressors, or an eductor or vacuum system can be used.

The volume of air displaced when transferring a product should also be considered. Without proper venting, the transfer operation will cease. If the transfer is very rapid, then correspondingly, the transfer vessel vent system must be capable of managing the "worst case" flow. Depending on the type of vapors released, the vent stream may have to be scrubbed, routed to the flare, or at least discharged at a safe location.

Stewart and McVey (1994) describe the design of a deinventory system for a hydrofluoric acid (HF) alkylation unit. The purpose of this deinventory system is to mitigate an accidental release by removing the hazardous materials rapidly from the leaking or damaged system to a safe location.

Secondary containment enclosures other than double-wall construction have been widely used to control vessel leaks. Bunkers built around underground storage tanks are common examples of this approach to preventing releases to the atmosphere and into the soil around the tank, to minimize potential for ground water contamination. Use of tank-high dikes with covers or roofs is another example of secondary containment that limits postrelease emissions.

5.6. Leak Plugging

While there are many techniques available to provide containment of leaks, often the only option available is to repair or reestablish the primary source of containment. This can be done by plugging, patching, or freezing, as described in the following sections.

5.6.1. Patching

Patching may be feasible for stopping leaks in vessels or storage containers, or in situations when the contents cannot be transferred. The usual patching method is to affix some type of mechanical device in, over, or around the leak to restrict flow. A variety of clamps, capping tools, and sleeves are commercially available for this purpose. For piping systems, one technique involves attaching a collar around the leak, which is subsequently filled with a patching material that sets in place. Several companies provide leak-patching equipment to handle various types of leaks. (A listing of leak repair specialists can be found in the annual Equipment Buyer's Guide issue of *Chemical Engineering* magazine.)

A special case involves the plugging of leaks in pressurized shipping containers. For example, The Chlorine Institute has developed a series of emergency kits designed to contain most leaks encountered in shipping containers, including ton cylinders, road tankers, and rail cars (The Chlorine Institute, 1969). Figure 5.4 shows this kit.

The decision to apply a mechanical constraint to stop a leak should be carefully weighed to ensure that the situation is not exacerbated. For example, a leaking relief valve on a rail car may be caused by overfilling or by a pressure-producing reaction. In such circumstances, it would not be appropriate to cap or gag the relief valve.

Personnel who may be called upon to use such devices risk bodily harm. Consequently, they should be properly trained in the use of plugging equipment, provided with proper personal protective gear, and trained in its use.

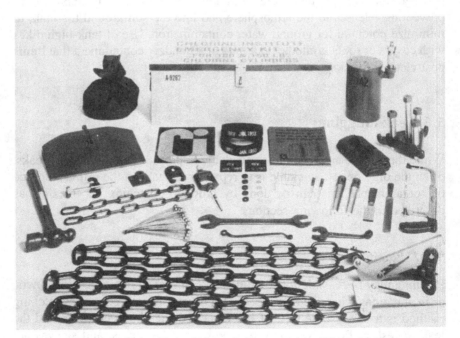

Figure 5.4. Emergency kit for chlorine leak repair. (Courtesy of the Chlorine Institute, Washington, D.C.)

5.6.2 Freezing

In a pipeline, small leaks may be stopped by applying external cooling to form a solid plug. The two most common sources of chilling are carbon dioxide fire extinguishers (–78°C) and liquid nitrogen (–196°C). A comparison of the freezing points of the leaking material and the cooling source will indicate if this approach is likely to be effective. The freezing points of many hazardous materials can be found in *Guidelines for Vapor Release Mitigation* (CCPS, 1988a).

When contemplating this approach, the construction materials must be carefully considered to determine whether low temperatures will cause them to become brittle or stressed. Furthermore, care should be taken to avoid asphyxiation when using large quantities of carbon dioxide, nitrogen, etc., especially in a confined space and inside buildings (Lees, 1980).

5.7. Physical Vapor Barriers

5.7.1. Overview

Vapor barriers can be considered containment devices as they are useful in retarding the dispersion of vapor clouds, and have already been alluded to in Section 5.2.1.

Recent work done on physical vapor barriers was performed by the Industrial Cooperative HF Mitigation/Assessment Program (ICHMAP) (Petersen and Radcliff, 1989). This program studied the effects of vapor fences and vapor boxes. The primary objective of the study was to determine the effectiveness of these devices to retard the transport and to dilute heavier-than-air (HTA) releases of a toxic material like hydrofluoric acid (HF). Because vapor barriers could also "see" releases of flammable materials, an effort was made to determine their impact on the consequences of a vapor cloud explosion.

5.7.2. Vapor Fences

Vapor fences are solid walls located downstream from an expected release point of a flammable or toxic gas. These fences are most effective in achieving an initial dilution when they are located as close to the source as possible but downwind of it, or on the side of a facility where a sensitive population is located. Heights of vapor fences can vary from 3 to 12 m. At times they are added to the tops of dikes to provide a dilution effect (see Section 5.2.1).

A vapor fence works by forcing toxic or flammable vapor to pile up behind it, then reach the fence height, and finally flow over the top of the fence. When the vapor begins to flow over the fence, it becomes entrained in the wind and dilution occurs. Until flow over the top occurs, the fence acts as a barrier with only small amounts of material being carried downwind.

Fences can assume various configurations. The simplest is a straight barrier. A more sophisticated straight vapor fence consists of several components of various lengths placed one behind the other with gaps between them. Another type of vapor fence forms a semicircle of a constant radius from the expected release point. Other configurations are possible and should be selected based on modeling studies (Meroney, 1991).

The interaction of dispersing clouds with vapor fences is a complex physical process. When a flow meets an obstruction, turbulence levels are increased downstream because of vorticities introduced into the flow field, and increased velocity gradients are induced by flow momentum losses. Detailed modeling of such a process is very difficult and requires a combination of small-scale experiments and computational fluid dynamics.

A simple concept can be used to illustrate important features of vapor barriers that alter cloud dispersion behavior in the near field (Meroney, 1991; Meroney and Neff, 1985). The concept is based on adding an entrainment velocity (u_e) contribution, which is attributed to the vapor fence.

Meroney (1991) and Meroney and Neff (1985) proposed a simple equation (5.1) to predict the increase in air entrainment by a vapor cloud caused by the presence of a vapor fence:

$$u_{e,vf} = C_D u_w(z_{vf})[1 - \rho]\left[1 - \frac{x - x_{vf}}{30 z_{vf}}\right] \tag{5.1}$$

where x is the distance downwind of the source, x_{vf} is the location of the vapor fence, z_{vf} is the vapor fence height, u_w is the wind speed evaluated at the fence height, ρ is the fence porosity, and C_D is the drag coefficient. Meroney (1991) and Meroney and Neff (1985) found good agreement between this simple equation and reported experimental data using a value of C_D of 1/10. Equation (5.1) is valid for distances less than or equal to x_{vf}.

Equation (5.1) indicates that dilution increases with fence height, since the wind velocity increases with elevation. Although equation (5.1) also indicates that increased wind speeds will result in larger entrainment rates, the cloud will pass through the fence more quickly. Therefore, the increased entrainment rates are offset, or compensated for, by the reduced residence time of the cloud in the wake and, as a net result, no net change in dilution rates is exhibited.

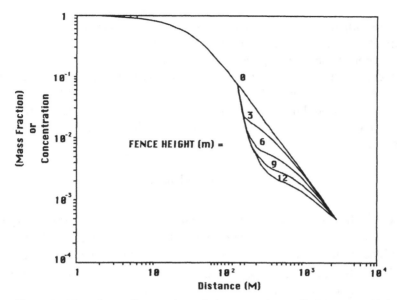

Figure 5.5. Vapor fence effects on downwind concentration profile (Meroney, 1991).

The additional entrainment contribution can easily be added to existing dispersion models, as illustrated by Meroney (1991) and Meroney and Neff (1985). Figure 5.5 shows the predictions of centerline concentrations of Goldfish Test 1 (Meroney, 1991) with different fence heights at a downwind location of 100 m. Note that the effects of initial dilution become negligible in the far field.

5.7.3. Vapor Boxes

A vapor box is very similar to a vapor fence, except that it is built around the source of the potential release and the process unit that would contain it. The ICHMAP study (Petersen and Radcliff, 1989) investigated two styles of boxes. The first was a straight-sided box with and without holes for doors cut in its sides. The holes were intended to represent openings necessary to gain access to the facility and over which it was thought water sprays could be placed to mitigate the effects of material that would leak through these openings.

Another type of box tested had flow spoilers mounted at the top, a configuration that would force the vapor back down into the box and toward the release points.

Assessing the effects of enclosures or vapor boxes is similar to assessing the effects of vapor fences. Consider the case of a typical dike sized to hold 110 percent of the liquid spilled. For small releases, the dike walls would act as containment or storage for the vapor generated shortly after release occurred. The vapor that would be generated by evaporation or boiling from the dike floor would displace the air in the dike for a given period of time and then overflow the dike walls. The vapor holdup duration is easily estimated:

$$A(z_d + z_f)\rho_v = \int_0^t \dot{m}_e \Delta t \tag{5.2}$$

where A is the dike area, z_d is the dike height designed to contain the liquid, z_f is the addition height for vapor containment, ρ_v is the vapor density of the spilled liquid at the pool temperature, $\dot{m}_e$ is the mass emission rate as a function of time, and Δt is the time it will take for vapor to be discharged from the vapor box (Arthur D. Little, Inc., 1974).

Equation (5.2) indicates that the time vapor is advected downwind will increase as z_f (height of vapor containment box) increases, as this increases the time to fill the vapor box. This additional time allows for the effects of decreasing rates of conductive heat transfer from the dike floor for cryogenic materials, or decreasing convective mass transfer for materials with boiling points that are higher than ambient temperatures, to take effect.

Equation (5.2) also indicates that the cloud's arrival time would be delayed at least by the time required for the vapor to overflow the vapor containment walls. Also, higher dilution in the near field should be exhibited, since the wind speed is higher at $z_d + z_f$ than at grade level.

Figure 5.6 illustrates the impact of vapor containment on downwind concentration levels based on scaled wind tunnel experiments (Van Zele and Diener, 1990). Far field reduction factors were observed and found to be dependent on the release rate, its duration and the barrier volume.

Although a vapor box or containment may be effective for concentration reduction, it may increase the explosion hazards. Higher explosion peak overpressure may be realized in the near field because of cloud geometry and partial blockage (Melhem and Croce, 1994).

Another interesting idea is to deplete vapor from the containment at a predetermined rate by using "leaky" or porous fence walls, as illustrated in Figure 5.7. The time required to reach z_d is

$$A\rho_v \frac{dz}{dt} = \dot{m}_e - \frac{2}{3}\rho_v C_D \sqrt{2g\left(\frac{\rho_v}{\rho_a} - 1\right)} z^{3/2} p\Pi \tag{5.3}$$

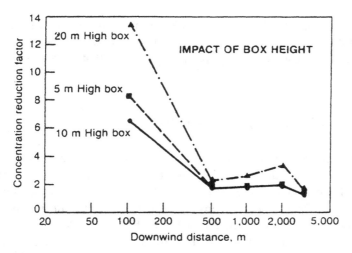

Figure 5.6. Vapor containment effects on downwind concentration profiles (Van Zele and Diener, 1990).

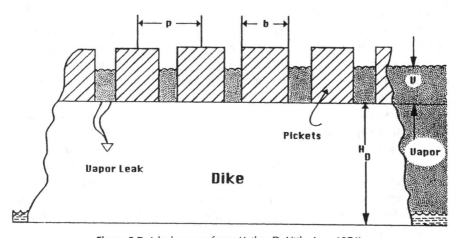

Figure 5.7. A leaky vapor fence (Arthur D. Little, Inc., 1974).

After t_d vapor begins to leak out. The height of the vapor will vary according to the following mass balance:

$$Az_d\rho_v = \int_0^{t_d} \dot{m}_e \, dt$$
(5.4)

where C_D is equal to $\frac{1}{2}$ for rectangular weirs, P is the porosity of vapor fence, and Π is the fence perimeter in the downwind direction. The second term on the right hand side of equation (5.4) represents the vapor leak rate. Arthur D. Little, Inc. (1974) showed that the reduction in the vapor evolution rate achieved by a leaky fence was not as effective as that for a solid fence.

5.7.4. Applicability of Vapor Barrier Devices

Van Zele and Diener (1990) state that vapor fences can reduce the near-field concentrations of hazardous materials, but that concentrations increase with increasing downwind distance. Wind tunnel data show that vapor fences will reduce near-field concentrations by factors of 2 to 9, but the concentrations downwind at approximately 1000 m from the release point will eventually equal concentrations when no fence is in place. For a heavier-than-air cloud, there is no appreciable delay in the cloud's arrival time.

Three factors strongly influence the effect any vapor barrier can have on cloud dilution. First, the release scenario, including the rate and duration of release, and the wind direction, will influence the barrier's effectiveness. Second, site layout, such as the location of process units and the equipment and buildings in them (e.g., tanks, towers, and pipe racks located upwind of the release point), can influence local air currents and the direction the release will take. Third, the height of the barrier and its relationship to other equipment and the source must be considered.

For cryogenics, vapor boxes may be more effective. Wind tunnel data indicate that vapor boxes have the potential to reduce near-field concentrations by a factor of 4 to 15 (Arthur D. Little, Inc., 1974). Additionally, far-field concentrations are lowered by the reduction in the rate the gas is discharged from the vapor box. This downwind reduction, when compared to a no-vapor-box case, can vary by a factor of 1 to 4. The degree of reduction is dependent on the release rate, the duration of the release, and the volume of the vapor box.

Testing, permitting, and inspection requirements also affect whether a vapor box is an appropriate choice. A vapor box is a confined space, which means that specific safety testing and permitting procedures must be followed. As a result, monitoring and servicing equipment inside the box may be difficult. Moreover, during operator rounds, the vapor box cuts off equipment from visual inspection. Also, the effect of a vapor box on flammable vapors may rule out its use. It has been found that vapor boxes increase peak explosion overpressures for flammable vapors by affecting cloud height. Providing explosion vents in the walls of the vapor box will somewhat reduce the peak explosion overpressure, but it will still be higher than if no vapor box were present. Lastly, decreasing the perimeter of the vapor box increases the peak overpressure from an explosion.

5.7.5. Effects of Process Equipment and Structures

Safety planners should keep in mind that the process equipment and structures upwind of the release point and vapor barrier can influence the performance of the vapor barrier. It is well recognized that these structures and equipment will affect the flow of the released gas or vapor around them and the degree of mixing that occurs.

A workshop report indicates that building architecture can induce a downwash of vapor clouds, cause channeling between structures, and produce local effects that could bring about mixing and other types of motion (Meroney, 1992). Therefore, emergency response plans for entry into process areas must anticipate the potential encounter of released vapors in high concentrations because of drafts and eddies caused by process equipment and buildings.

5.8. References

API Standard 620. Rev. 1, 1985. *Design and Construction of Large, Welded, Low-Pressure, Storage Tanks,* Appendices Q and R, 7th ed. 1985. Washington, D.C: American Petroleum Institute,.

Arthur D. Little, Inc. 1974. Evaluation of LNG Vapor Control Methods. Report to the American Gas Association. Cambridge, MA: Arthur D. Little, Inc.

Bauer, W.H., D. N. Borton, and J. J. Bulloff. 1975. "Agents, Methods and Devices for Amelioration of Discharges of Hazardous Chemicals on Water." Report prepared for the U.S. Department of Transportation, Contract no. DOT-CG-42759-1.

29 C.F.R. 1910.146. 1993. Occupational Safety and Health Administration.

32 C.F.R. Part 650.1992. Environmental Protection and Enhancement Act.

40 C.F.R. Part 267. 1990. Surface Impoundments. Environmental Protection Agency, (EPA).

49 C.F.R. Part 193. 1990. LNG: Federal Safety Standards. Department of Transportation.

CCPS (Center for Chemical Process Safety). 1988a. *Guidelines for Vapor Release Mitigation.* New York: American Institute of Chemical Engineers.

CCPS (Center for Chemical Process Safety). 1988b. *Guidelines for Safe Storage and Handling of High Toxic Hazard Materials.* New York: American Institute of Chemical Engineers.

Chemical Industry Association. 1975. *Code of Practice for Large-Scale Storage of Fully Refrigerated Anhydrous Ammonia in the United Kingdom.* London: Chemical Industry Association.

Chlorine Institute. 1969. *Chlorine Manual.* 4th ed. Washington D.C. The Chlorine Institute Inc.

Dilwali, K. M., and K. S. Mudan. 1987. Dike Design Alternatives to Reduce Toxic Vapor Dispersion Hazards. Paper presented at the *AIChE Vapor Cloud Conference Proceedings*, Boston, MA. New York: American Institute of Chemical Engineers.

Drake, E.M., and R. C. Reid. 1975. How LNG Boils on Soils. *Hydrocarbon Processing* (May): 191–194.

Feind, K. 1975. Reducing Vapor Loss in Ammonia Tank Spills. *Proceedings of the Safety in Ammonia Plants and Related Facilities Symposium,* Salt Lake, pp. 114–118. New York: American Institute of Chemical Engineers

GRI (Gas Research Institute). 1990. "Development of Polymer Concrete for Dike Insulation at LNG Facilities; Phase IV, Low-Cost Materials." Final Report no. GRI 90/0259. Chicago, IL: Gas Research Institute.

Hangs, F. E. 1986. Hand-held Calculator Simplifies Dike Computations. *Pipe Line Industry* 65(2): 32–33.

Harris, G. E., D. S. Davis, G. B. DeWolf, and J. D. Quass. 1987.Control of Accidental Releases of Hydrogen Fluoride. Paper presented at *Symposium on Safety in Chemical Operations.* American Institute of Chemical Engineers Annual Meeting. New York.

Harris, N. C. 1991. Containment Building and Scrubber for Toxic Gas Plants. *Proceedings of the International Conference and Workshop on Modeling and Mitigating the Consequences of Accidental Releases of Hazardous Materials*, pp. 443–452. New York: American Institute of Chemical Engineers.

Lees, F. P. 1980. *Loss Prevention in the Process Industries.* 2 vols. London and Boston: Butterworths.

Mazzarella, M. M. 1984. Economical Proportioning of Spill-Retention Dikes. *Plant Engineering.* (December 13): 60–62.

Melhem, G. A., and P. A. Croce. 1994. "Advanced Consequence Analysis: Emission, Dispersion, Fire and Explosion Dynamics." Working Manuscript, Arthur D. Little, Inc.

Meroney, R. N. 1991. Numerical Simulation of the Mitigation of HF Cloud Concentrations by Means of Vapor Barriers and Water Spray Curtains. *Proceedings of the International Conference and Workshop on Modeling and Mitigating the Consequences of Accidental Releases of Hazardous Materials*, pp. 431–442. New York: American Institute of Chemical Engineers.

Meroney, R. N. 1992 Dispersion in Non-Flat Obstructed Terrain and Advanced Modeling Techniques. *Plant Operations Progress*, 11(1): 6–11.

Meroney, R. N., and D. E. Neff. 1985. Numerical Modeling of Water Spray Barriers for Dispersing Dense Gases. *Boundary Layer Meteorology* 31: 233–247.

Moorhouse, J., and R. J. Carpenter. 1986, "Factors Affecting Vapor Evolution Rates from Liquefied Gas Spills." British Gas Corporation, Midlands Research Station.

NFPA 30. 1990. Flammable and Combustible Liquid Codes. Quincy, MA: National Fire Protection Association.

NFPA 68. 1994 Edition. *Guide for Venting of Deflagrations.* Quincy, MA: National Fire Protection Association.

Petersen, R. L. and M. A. Radcliff. 1989. "Vapor Barrier Assessment Program for Delaying and Diluting HF Vapor Clouds." Report prepared on behalf of Joint Industry Cooperative HF Mitigation/Ambient Assessment Program.

Plastics World. 1987. "New Chemsorb™ Universal Sorbent Reduces Threat of Hazardous Liquid Cleanup." New Release no. 1 (24 Nov.).

Prugh, R. W. 1992. Pre-Release Mitigation. *Plant/Operations Progress.* 11(1): 19–22.

Stewart, N. E. and J. W. McVey. 1994. Design Concepts for a Hydrogen Fluoride Emergency De-Inventory System. *Process Safety Progress* 13(2): 105–107.

Van Zele, R. L. and R. Diener. 1990. On the Road to HF Mitigation—Part 2. *Hydrocarbon Processing* (July): 77.

Welker, J. R. 1987. "Thermal Radiation and Vapor Dispersion Exclusion Zones." Paper presented at the *ASME Pipeline Engineering Symposium*. Dallas, TX. New York: American Society of Mechanical Engineers.

Welker, J.R., and D. W. Johnson. 1987. Diked-in Storage Areas, Revisited. *Chemical Engineering* (July): 112.

Zanker, A., 1979. Nomograph Quickly Finds Tank's Diked-in Area. *Process Engineering* (May): 67–69.

6

Detection and Response

6.1. Introduction

Chapter 6 addresses issues associated with postrelease mitigation, namely, the detection of a leak and the response of an organization to the event. Chapters 3, 4, and 5 cited numerous needs to detect releases and activate postrelease mitigation systems. In this chapter, the hardware that is available to perform this function, their operating principles, and maintenance requirements to ensure maximum system reliability are reviewed in some detail.

Following a release, the plant staff must respond and alert the surrounding population that an emergency exists on the site. This is facilitated by having a well-thought-out and rehearsed emergency response plan in place. The essential elements of this plan are associated with postrelease mitigation.

Because of the rapidity with which a release can develop, there may not be adequate time to conduct an emergency evacuation of populations close to the site, nor of nonemergency response personnel on the site. Instead they may have to be sheltered in place. The requirements for setting up shelters for this purpose and evaluating their effectiveness is also described in this chapter.

6.2. Leak Detection

The need for devices to detect leaks has been discussed in earlier chapters. The reason for their use is that they can be permanently located in an area, or at a piece of equipment to provide an early warning of a leak. At the same time there are many types of portable instruments that can be used to detect leaks and warn personnel in those areas where hazardous materials are concentrated.

Leak detectors may be divided into two general categories, fixed systems and portable systems, as illustrated in Figure 6.1.

As shown in the figure, fixed systems can be broken down into two general categories: those with sensors in place, and those that monitor an area or location by drawing a sample to a central analyzer. Both of these categories can be further broken down into systems for dealing with flammable or toxic gases. Portable instruments break down into similar categories.

Since each sensor type may be used in different detection classifications, the following discussion focuses on the different types of sensors and the bases of their design and operation.

6.2.1. Fixed-Point Detectors

Fixed-point detectors usually feature single detector heads that have been permanently mounted at a specific location to detect leaks of toxic or flammable gases. The instruments are usually connected by a three-wire cable with one wire providing power to the detector, one the signal from the detector, and one serving as a system ground wire. The signal from the detector is fed back to a central alarm panel located in a control room or security office where, if a high signal level is detected, an alarm or automatic mitigation action (such as closing isolation valves) is initiated. Since each detector has a specific location, it is easy to pinpoint the leak area. The different technologies used for fixed-point detectors are described in the following sections.

6.2.1.1 Catalytic Sensors

The operation of a catalytic sensor depends on the oxidation of the flammable gas on the detector which is an electrically heated catalytic filament. This detector is usually made of fine platinum wire that is coated with one of several alternative substances—palladium/platinum/thoria catalysts. These catalysts help to enhance the catalytic activity and prolong the life of the detector by enabling it to operate at lower temperatures. The detector filament is connected to an identical, but inactive, unit in a Wheatstone bridge, and is located adjacent to the active detector. The identical inactive unit allows for ambient temperature compensation.

When a flammable gas enters a catalytic cell, the gases are combusted on the active filament, causing an increase in temperature and a change in the balance of the Wheatstone bridge. The output signal for flammable gases in air in this type of cell is linear up to the lower explosive limit (LEL).

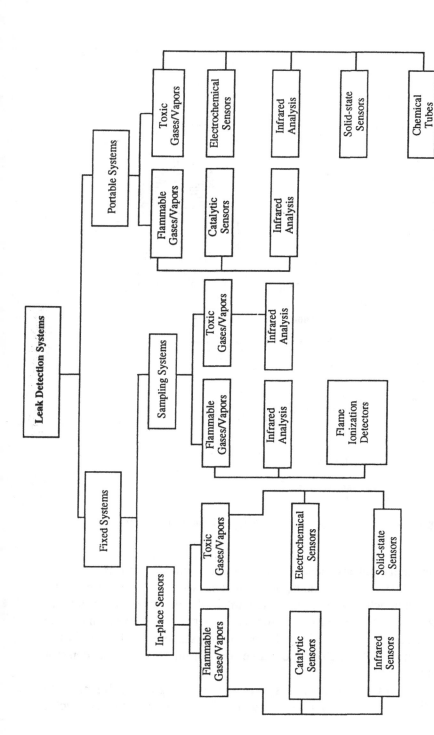

Figure 6.1. General Breakdown of Leak Detection Equipment

117

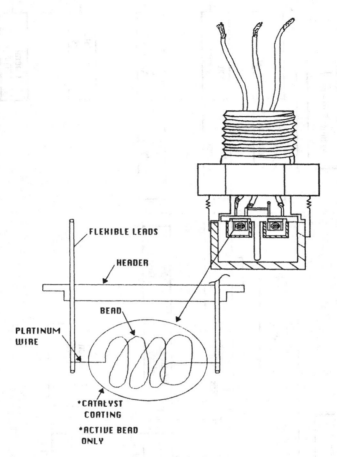

Figure 6.2. Sketch of catalytic sensor. (Courtesy of Sensidyne Inc., Clearwater, FL.)

Many industrial atmospheres contain contaminants such as silicones, organic lead, sulfur compounds, and halogens which can poison catalytic sensors. Manufacturers go to great lengths in the design of these units to minimize loss of sensor activity from such influences.[1] Means taken to keep water, corrosive liquids, and dirt from entering sensors involve the installation of membranes before the flame arrestor in front of the detector filament.[2]

The reason that manufacturers of these detectors go to such lengths to protect the instruments from poisoning and burnout is that a failed detector

1 Product brochure. Sensidyne Inc., Clearwater, FL.
2 Product brochure. Teledyne Analytical Instruments, City of Industry, CA.

will not register a voltage difference, so will not indicate failure. As a result, the potential exists for a leak to go undetected. Therefore, when using these catalytic sensors, system maintenance and calibration are critical. Figure 6.2 is a sketch of a catalytic sensor element.[3]

6.2.1.2 Infrared Detectors

Infrared detectors are used to detect the presence of flammable gases. Their principle of operation is based on tuning the infrared (IR) light beam to the hydrogen carbon bond (C—H) that is found in flammable materials (see Figure 6.3). The flammable gas that passes through this beam absorbs some of the IR light, causing a change in transmission of the light. This anomaly is picked up by the detector and converted to a standard 4- to 20-milliAmp output signal. Infrared detectors are not subject to any poisons, but if moisture condenses on the instrument's reflecting mirror it will absorb the

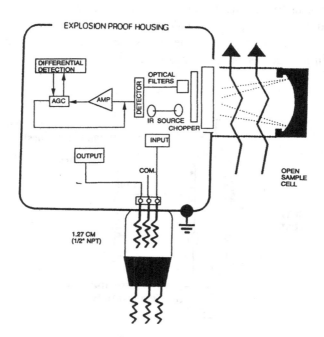

Figure 6.3. Schematic of an IR detector for flammable gas detection.
(Courtesy of Astro International Corp.)

---

3 Product brochure. Sensidyne Inc., Clearwater, FL.

IR beam and cause a false alarm. This problem is overcome by installing heaters to keep the mirror area warm and prevent condensation.[4]

In addition, the application of these sensors to the monitoring of carbon dioxide concentrations is also reported. Figure 6.3 shows the cross-section of an IR detector.

6.2.1.3. Solid-State Sensors

The two key elements of a solid-state sensor (Figure 6.4) are the heater and collector electrodes that are imbedded in a chip of solid-state material that has been doped with a metal oxide, or a mixture of metal oxides, to make it responsive to a specific gas or group of gases. A constant current is passed through the heater electrode to keep the surface of the sensor at a constant temperature. Upon exposure to the toxic or flammable gas being monitored, the surface resistance changes. This change is picked up by the detector electrode, then processed and converted to an appropriate output signal.

By using different metal oxides on the surface, solid-state detectors can be made sensitive to many flammable and toxic gases. Also, they are not poisoned by silicones, lead, or halogens. As their response does not require oxygen to operate, solid-state detectors can be used to detect hazardous vapors in nitrogen or helium atmospheres.[5]

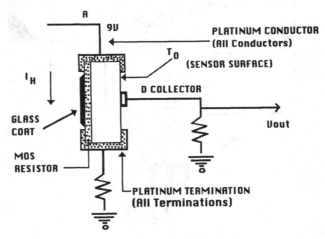

Figure 6.4. Schematic of a solid-state detector.
(Courtesy of International Sensor Technology, Irvine, CA.)

---

4 Product brochure. Astro International Corp., League City, TX.
5 Product bulletin. International Sensor Technology, Irvine, CA.

6.2.1.4. Electrochemical Sensors

Electrochemical sensors (Figure 6.5) are useful for detecting toxic gases. Their principle of operation is based on an electrochemical reaction of the gas with an electrolyte contained in the sensor. This is a reduction–oxidation reaction in which electrons are generated in proportion to the concentration of the gas present. This electron flow is then electronically processed to give an output signal. In many cases this output signal is linear with gas concentration. An example is the use of electrochemical sensors to measure hydrogen sulfide.[6] The hydrogen sulfide enters the sensor through a diffusion barrier and gas-permeable membrane. A reaction takes place at the working electrode (anode) releasing electrons that flow to the counter-electrode (cathode) where a second reaction occurs. These reactions may

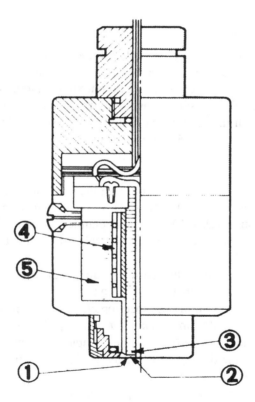

Figure 6.5. Cross-sectional view of an electrochemical sensor. *Key:* 1. Membrane; 2. Thin-film electrolyte; 3. Working electrode; 4. Counter electrode; 5. Electrolyte. (Courtesy of Enmet Corporation, Ann Arbor, MI.)

6 Product brochure. Lumidor Safety Products, Miramar, FL.

or may not consume the electrolyte contained in the sensor. The reactions for hydrogen sulfide detection are shown below.

$$\text{Anode reaction: } H_2S + 4H_2O \rightarrow H_2SO_4 + 8H^+ + 8e^- \qquad (6.1)$$

$$\text{Cathode reaction: } 2O_2 + 8H^+ + 8e^- \rightarrow 4H_2O . \qquad (6.2)$$

Because their electrolytes can be consumed or products of the reactions can build up in them, the sensors are designed to be easily refilled or to be disposed of and replaced in the field.[7]

Electrochemical sensors are applicable to many toxic gases and are useful for the detection of hydrogen, which is not responsive to the catalytic sensors described earlier.

6.2.2. Sampling Systems

Sampling systems are centrally located leak-detection systems that use a pump or aspirator to pull samples from various locations into an instrument for analysis. They can be used to detect both toxic or flammable materials. The instruments used may sample and analyze a single point or multiple points, depending on the application. Additionally, some of these systems will allow detection of up to five components.

6.2.2.1. Infrared (IR) Analyzers

The IR detection method will work for more than 300 of the approximately 500 gases and vapors declared hazardous by the Occupational Safety and Health Administration (OSHA). It is also claimed to detect and measure many of the 189 hazardous air pollutants defined by the Environmental Protection Agency (EPA).[8]

Remote multicomponent air samples can be drawn into a centrally located analyzer (under computer control) and then into the gas cell of an infrared spectrometer. Within the spectrometer, a system of lenses and mirrors passes an infrared beam in a predetermined path through the sample. The amount of energy absorbed by the sample is compared against a standard beam, and the difference is related to the concentration of the gas of interest. By changing the wavelength of the infrared beam, additional materials may be checked for in the gas sample and their concentration levels determined the same way. If the concentration of the compound of interest exceeds a predetermined level, an alarm is activated.

---

7 Product brochure. Sensidyne Inc., Clearwater, FL.
8 Product brochure. The Foxboro Company EMV, East Bridgewater, MA.

In addition to the above, the data on the actual measurements can be stored and used to generate reports on concentrations of the compounds of interest.

Because of the machine's mechanical limitations, there is a limit of five components that can be analyzed at one time at about 24 remote sampling points.[9]

By tuning these instruments to a single IR wavelength, one compound can be monitored for continuously at multiple sample locations. This mode of operation is similar to the single-head IR detector described in Section 6.2.1.2.

6.2.2.2. Flame Ionization Detectors
The flame ionization detector (Figure 6.6) is also known as a total hydrocarbon analyzer and is used for detecting flammable hydrocarbon vapors.

In its use, samples are drawn into the instrument through the ionizing cell by a pumping system. In the ionizing cell there is a hydrogen flame that causes the hydrocarbons in the gas stream to break down into charged ions. The positive ions are attracted to an oppositely charged, high-voltage collector, creating a current flow proportional to the number of carbon atoms in the sample. This current is measured by an electrometer and transmitted to the instrument monitor where it is displayed as a reading.[10]

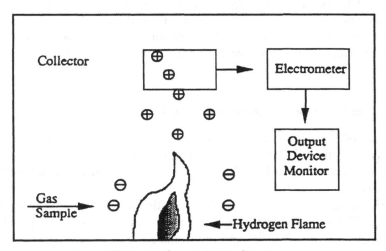

Figure 6.6. Flame ionization detector.

9 Product brochure. Control Instruments Corporation, Fairfield, NJ.
10 Product brochure. Control Instruments Corporation, Fairfield, NJ.

As in the case of the infrared instrument described in the preceding subsection, multiple points can be hooked to one analyzer and sampled periodically.[11]

The detectors and their wiring must meet the requirement of the area electrical classification in which they are being installed.

6.2.3. Portable Detectors

6.2.3.1. Overview

Portable detectors for toxic or flammable gases are essential for use by an emergency response team so that they will know the hazards they are facing when responding to a release. There are many portable analyzers that utilize the fixed-point and sampling system technology described in Sections 6.2.1 and 6.2.2. One other method of portable detection is the use of chemical tubes.

6.2.3.2. Chemical Tubes

Chemical tube detectors are chemical-specific and require the use of calibrated tubes for each vapor to be detected. The ends of the tube are broken off and placed in a pump that draws a fixed sample volume through the tube. The gas sample undergoes a colorimetric reaction with the tube packing. The greater the concentration of the sought-for chemical, the greater the amount of the material in the tube that will be reacted. The concentration of the gas can then be determined by reading the length of the tube that is colored against a concentration scale printed on the side of the tube. Tubes of this type are available for more than 350 different gases and vapors.[12]

6.2.4. Detector System Response Times

For the systems described in the previous sections, there is a wide variation in response times. As some of the examples in Chapter 7 will show, quick detection of a release and timely activation of postrelease mitigation systems is essential.

Fixed-point sensors, such as the catalytic, electrochemical, or infrared types, will respond rapidly to gas exposure and generate an output signal within 10 to 30 seconds of exposure to the gas.[13]

---

11 Product bulletin. Control Instruments Corporation, Fairfield, NJ.
12 Product brochure. National Draeger Inc., Pittsburgh, PA.
13 Product brochures. Sensidyne Corporation, Clearwater, FL, and Control Instruments Corporation, Fairfield, NJ.

For sampling systems that look at a single or multiple points, the response time can be much longer. The exact time will depend on the time needed for the instrument to completely analyze the sample, and the time needed to draw a sample through the tubing to the instrument. Sampling time must be long enough to ensure that a current sample is obtained and analyzed, rather than reanalyzing vapor from the previous sample. Obviously the more remote the sampling point from the instrument, the longer the time needed to get a true sample and the higher the gas flow rate in the sampling system.

If the sampling system is looking at multiple points, then each point will be looked at only once in the total time per sample analysis, times the number of sample points being monitored.

For portable sampling systems, the response time should be quite short because the sample is being taken from the surrounding environment, or from a nearby location such as inside a confined space. Portable instruments with internal pumps can provide continuous readings of vapor concentrations. Their use is limited only by the life of the battery pack.

6.2.5. Detector Placement

The following are some general guidelines on the placement of gas detectors. The manufacturers of detectors should be consulted for recommendations on placement of their detectors, since they are more familiar with their performance characteristics and capabilities. At the same time the system designer must do his/her own research to ensure the equipment being installed will deliver the required results.

In general, single-point detectors, such as the catalytic, infrared, solid-state, and electrochemical types, can be put in locations up to 3,000 feet from the control room.[14] Sampling systems are more limited because of the length of the tubing needed to bring the sample to the detector.

Generally, when considering the placing of detectors, one should consider the sources around the site that are most likely to lead to a release such as:

- pump seals;
- compressors;
- analyzer shacks;
- storage tanks or vessels;
- cooling towers above the fans;
- air-cooled heat exchangers; and
- unloading or loading areas.

14 Product brochure. Sensidyne Corporation, Clearwater, FL.

The second tier of locations for detectors are those areas that could be affected by a release, such as:

- air inlets to control rooms;
- locations downwind of process units that are populated (e.g., administration buildings, maintenance shops, neighboring communities, etc.);
- the fence line, for off-site concerns or to warn of a release from a neighboring facility;
- the area between ignition points (furnaces, boilers, flares, etc.) and the potential process release point; and
- air intakes for buildings that are to serve as temporary shelters in the event of a release.

Also regarding location of detectors, the elevation at which they are placed must take into account the vapor's density. Detectors for dense gases, like chlorine, are typically mounted at a height of from 0.3 to 1 m, while for lighter gases, detectors are placed in the breathing range at 1.5 to 2.4 m above grade. Detectors placed at the fenceline may not detect elevated releases leaving the site, so elevated detectors may also be required. Lastly, the designer of the system must understand equipment limitations to assure selection of proper equipment to meet the specified needs.

6.2.6. System Reliability

6.2.6.1. Fixed Systems
For a leak-detection system that features fixed-point detectors to be reliable, it must be monitored and calibrated frequently. As discussed in Section 6.2.1.1, catalytic detectors for flammable gases can either burn out or be poisoned by other materials in the atmosphere. When burnout or poisoning occurs, the detector still behaves as if it is working properly. Therefore, the only way to detect failure is by frequent calibration checks. A similar situation exists for solid-state and electrochemical detectors. For these devices, manufacturers recommend that calibration checks be made at frequent intervals.

Calibration checks also serve to minimize nuisance alarms and false trips of postrelease mitigation systems. This ensures that the personnel monitoring the system will have confidence in it and will respond aggressively when an alarm is received. If there are many nuisance alarms or false trips of postrelease mitigation systems, it is highly likely that an alarm will be ignored and the system turned off, or the response time will be slow because someone will go to the location to determine whether there is really a problem. In the case of an actual release, time will have been lost in activating the postrelease mitigation systems and the emergency response plan.

Manufacturers of fixed-point leak detection sensors can furnish all the equipment and gases required to perform calibration checks.

6.2.6.2. Portable Gas Detectors

Portable gas detectors, as mentioned earlier, are generally used for two purposes: for measuring atmospheric contamination in confined spaces; and for checking the atmosphere that emergency response teams are entering or in which they are working. In the first case the portable instrument must be calibrated and checked by a trained individual (29 C.F.R. sec. 1910.146 App. C). In the emergency response case, the instrument must be ready for instantaneous use. In this case a higher and more frequent standard for calibration and maintenance is needed. Manufacturers of portable instruments can provide the necessary calibration equipment and service for their instruments.

6.2.6.3. Management of Change

The integrity of any process, whether it be a mitigation system, a leak-detection system, or the main process from which a release is generated, can be easily compromised with the smallest addition or modification. Actions as simple as bypassing an alarm or adding a block valve can quickly keep a system from operating when called upon.

Therefore, if a leak-detection system has been installed to warn of releases of toxic or flammable materials from a plant, the implementation of a process safety management system should include the detection system as part of the process (CCPS 1989; CCPS 1988). In this manner the leak-detection system will be subject to the facility's management-of-change procedures, and changes to the leak-detection system, will be carefully evaluated before being implemented.

6.3. Emergency Response

6.3.1. Introduction

The requirements of an emergency response plan and what it should contain are detailed in several OSHA regulations. Also, the proposed risk management plan regulations of the EPA (40 CFR, Part 68) contain some specific requirements that must be met by an emergency response plan (29 C.F.R. sec. 1910.38; 29 C.F.R. sec. 1910.119; 29 C.F.R. sec. 1910.120). In formulating any emergency response plan, it is important that it be made to meet the requirements spelled out in these regulations.

6.3.2. Fundamentals of a Comprehensive Emergency Response Plan

Table 6.1 lists general areas that should be addressed in a comprehensive emergency response plan. Specific tailoring is recommended, based on the particular facility, its operations, its storage and use of hazardous materials, contiguous land use patterns, support from public authorities, and other considerations.

The plan should be administered on a continued basis, with particular attention paid to its effectiveness during second and third shifts, and holidays, in light of the reduced number of personnel on-site at these times. The plan should be done in concert with emergency shutdown or other operational needs.

It is also critical to maintain the emergency response plan over time. Changes in operations should be reflected, whether they are process-, equipment-, or personnel-related. Similarly, training should be conducted periodically and emergency response equipment maintained regularly as part of a formal preventive maintenance program.

6.3.3. Emergency Response Training

6.3.3.1. On-Site Population

On-site population includes not only employees (with specific training for the emergency response team), but also contractors and visitors. Every person who enters a facility should be trained in how to respond to an emergency. In most cases, this involves a simple orientation process to familiarize everyone with the basic hazards that exist at the facility, general emergency recognition, and the facility's responses to alarms. Personnel who work closely with the process and may be instrumental in its control will certainly receive more training as their roles necessitate.

Personnel who enter the facility on a day-to-day basis receive training in a number of ways, but often the infrequent or nonindustrial visitor is overlooked and receives no emergency training. Such visitors might include:

- Tank truck drivers
- Sanitation truck drivers

- Package delivery people
- Pest control personnel
- Copier repairmen
- Evening janitorial staff
- Temporary secretarial support
- Building service repairmen (e.g., plumbers, carpet layers, HVAC repairmen)
- Railroad switch crews
- Salesmen

All personnel working within a facility should always be trained to recognize and respond to an emergency, including instruction in how to notify appropriate plant officials and what actions to take in case an evacuation is required (29 C.F.R. sec. 1910.38), (CCPS, 1995).

6.3.3.2. Off-Site Population

Emergency response training of the off-site population is critical in ensuring that, if a release does reach beyond the facility's fence line, harm to the general public will be minimized. As a minimum, the facility should inform the local emergency planning committee (LEPC) and the fire and police departments of the potential hazards that the facility could generate and then, as a group, they should develop effective response measures. These measures could include:

- Community evacuation
- Sheltering in place
- Blocking of highways
- Restriction of flight paths
- Evacuation points and routes
- Equipment staging points
- Incident command locations
- Special response equipment
- Medical procedures anticipated to diminish the effects of exposures to releases.

Training of the local populace can best be initiated through a community awareness campaign in which the community can become familiar with the facility, its processes and the potential hazards they present, and plant personnel with whom they would interact during an emergency.

TABLE 6.1

Contents for a Comprehensive Emergency Response Plan

Preface to Plan   • List of Key Telephone Numbers   • Table of Contents   • Preface   • Record of Changes   • Distribution List
Corporate Policy and Plan Objectives   • Introduction   • Purpose of Plan   • Corporate Policy Regarding Emergency Planning   • Coordination and Cooperation with Local Emergency Planning Committees   • Corporate Community Awareness Program   • Update Procedures and Schedule

Emergency Response Organization: Structure and Duties
 • Authority
 • Designation of Site Emergency Coordinator/Commander and Alternates
 • Duties of Site Emergency Coordinator/Commander
 • Chain of Command
 • Designation of Key Personnel and Alternates, including
   —Medical Emergency Director
   —Fire Brigade Team Leader
   —Oil Spill Response Team Leader
   —HazMat Spill Response Team Leader
   —Public Affairs Coordinator/Press Officer
   —Legal Counsel
   —Security Coordinator
   —Insurance and Claims Coordinator
   —Emergency Communications System Coordinator
   —Coordinator of Personnel Services
   —Purchasing and Logistical Support Coordinator (including transportation, supplies, etc.)
   —Maintenance and Engineering Support Coordinator
   —Scientific/Environmental Support Coordinator
   —Record-keeping and Documentation Coordinator
   —Others (as needed)
 • Responsibilities and Duties of Key Personnel
 • Designation of Facility Emergency Operations Center
 • Establishment of Field Emergency Command Post

Detection, Alarm, and Notification Procedures
 • Notification Check-off Lists for Various Types of Events
 • Telephone Rosters
 • On-site Accident Detection and Alarm Procedures
 • Identification and Policy for Notification of
   —Local Authorities
   —State Authorities
   —Federal Authorities

—Various Facility Personnel
—Special Off-site Occupancies
—Downstream Water-Users
—Water Treatment Plants
—Electric and Gas Utilities
—Air, Rail, and Marine Traffic
—Corporate Management
—Person(s) Responsible for Initial Notification and Continuation of Communications (where necessary)
—Documentation of Communications during Emergencies

Emergency Communication Systems
- Intrafacility Communication Systems
- Links to Public Authorities
- Need and Availability of Backup Systems
- Links to the Public

On-Site Evacuation and Security
- On-site Alert/Alarm Systems
- Employee Awareness Program
- Evacuation Procedures and Routes
- Provisions for Handicapped Employees or Visitors
- Places of Refuge
- Designated Assembly Areas and Alternates
- Procedure for Identification of the Missing
- Search and Rescue Procedures
- On-site Traffic Control
- Accident Site Boundary Control
- Facility Access Control
- Procedures for Protection of Vital Records
- Periodic Drills
- Documentation of Drills and Activities During Emergencies

Emergency Facility Shutdown Procedures
- References to Unit Emergency Operating Procedures
- Operator Training for Emergencies
- Safety Features of Control Rooms and Other Key Locations for Control of Facility Processes
- Drills and Exercises
- Documentation of Training and Drills

Medical Emergency Preparedness
- On-site Health Care Resources
- Available Ambulance Services and EMTs
- Coordination of Emergency Planning with Off-site Medical Facilities
- Toxic Substance Information Resources
- Special Antidotes and Supplies
- Victim Decontamination Procedures
- Provisions to Protect Medical Providers from Contamination
- Documentation of Activities before and during Emergencies

(Table continues on next page)

TABLE 6.1 (continued)
Contents for a Comprehensive Emergency Response Plan

On-Site Emergency Response Teams
- Fire Brigade Team Organization
- Oil Spill Team Organization
- HazMat Team Organization
- Response Team Activation Procedures
- Response Times during Normal Working Hours
- Response Times on Nights and Weekends
- State and Federal Training Requirements
- Actual Training Provided
- Drills and Exercises
- Documentation of Training and Drills

Personal Protection of Response Teams
- List of Hazardous Materials Likely to Be Encountered
- Selection of Respiratory Protective Devices
- Availability of Respiratory Protective Devices
- Resupply of Self-contained Breathing Apparatus
- Selection of Protective Clothing
- Availability of Protective Clothing
- Medical Surveillance and Care
- Decontamination Procedures
- Documentation of Activities during Emergencies

Fire Response Procedures
- General Procedures by Incident Type
- Required Resources for Postulated Events
- Available Resources for Postulated Events
- Documentation of Activities during Emergencies

Spill Containment and Cleanup Procedures
- Plugging/Stopping of Leaks
- Suppression of Hazardous Gas or Vapor Evolution
- Intentional Ignition of Combustible Toxic Gases
- Containment of Spills on Land
- Cleanup of Spills on Land
- Containment of Spills on Water
- Cleanup of Spills on Water
- Support Services for Response Forces
- Field Maintenance of Response Equipment
- Waste Handling and Disposal
- Documentation of Activities during Emergencies

Environmental and Spill Monitoring
- Tracking of Oil Spills on Water
- Surveillance of Other Types of Spills
- Monitoring of Atmospheric and Environmental Conditions
- Sampling/Monitoring of Environmental Contamination
- Documentation of Activities during Emergencies

Public Relations in Emergencies
- Corporate Policy Directives
- Designation of Media Briefing Location
- Communications with the Command Center
- Provision of Press Officer Support (including legal counsel)
- Provision of Support to Media Personnel
- Coordination and Cooperation with Public Authorities
- Availability of Press Kits
- Media Contact List
- Documentation of Activities and Statements during Emergencies

Application of Plan to Natural Hazards
- Procedures and Policies for Various Hazards
- Need and Availability of Special Resources
- Documentation of Activities during Emergencies
- Documentation of Activities and Statements during Emergencies

Off-site Post-Incident Recovery
- Status Monitoring of Injured Parties
- Wildlife Rescue and Rehabilitation
- Assessment of Environmental Damage
- Restoration of the Environment
- Assessment of Off-site Property Damage or Loss of Business
- Availability of Insurance and Other Resources
- Claim-handling Procedures
- Documentation of Activities

On-site Post-Incident Recovery
- Post-incident Response Debriefing and Review
- Assistance to Families of Injured Employees
- Assistance to Employees Laid-off Due to the Incident
- Investigation of Causal Factors
- Site Property Damage Assessment
- Site Decontamination and Cleanup
- Waste and Debris Disposal
- Site Reconstruction/Restoration
- Post-incident Response Debriefing
- Pre-startup Checklist
- Documentation of Activities

Off-site Sources of Assistance
- Corporate Resources
- Mutual-Aid Cooperatives
- Spill Cleanup Contractors
- Other Types of Key Contractors
- Expert Consultants
- Local Government
- State Government
- Federal Government
- Private and Volunteer Organizations
- Expected Response Time

(Table continues on next page)

TABLE 6.1 (continued)
Contents for a Comprehensive Emergency Response Plan

Resource Listings: Supplies and Supplemental Services
- Specialized Medical Supplies
- Oil Spill Dispersants and Application Equipment
- Oil Spill Burning/Wicking Agents and Application Equipment
- Sorbent Materials
- Neutralization Agents
- Portable Liquid Transfer Systems
- Temporary Storage Containers and Systems
- Vacuum Trucks
- Fire Extinguishing Agents and Equipment (including foams for vapor suppression)
- Portable Contaminant Detectors/Monitors
- Contaminant Sampling Equipment and Supplies
- Laboratory Analysis Services
- Communications Equipment
- Earth-moving Equipment
- Oil Skimmers
- Support Boats
- Fixed-wing Aircraft
- Helicopters
- Trucks, Vans, and Buses
- Emergency Lighting
- Emergency Power Generators
- Fuel Supplies
- Canteen Services
- Temporary Housing
- Portable Sanitation Facilities
- Work Clothes and Footwear
- Photography and Videotaping Services
- Sources of General Hardware
- Other Potentially Required Supplies and Equipment Not Addressed Here or in Previous Sections

---

Facility Planning Basis and Hazard Analysis
- Facility Layout and Maps
- Details of Potentially Hazardous Operations
- Oil/Chemical Accident Prevention Measures
- History of Past Accidents/Incidents
- Type and Expected Frequency of Natural Hazards
- Potential Impact of Natural Hazards
- Credible Accident Scenarios and Associated Probabilities
- Estimated Accident Impacts and Associated Safety Zones
- Identification and Description of Environmentally Sensitive Areas Subject to Impact
- Procedures for Real-time Hazard Assessment During Emergencies
- Technical References

Source: Arthur D. Little, Inc.

6.4. Community Relationships and Interactions

A facility's interaction with the community regarding emergency response is important in two general areas. First, there is need to coordinate the public response to an emergency with that of the facility. Relationships with the local fire and police departments, rescue squad, and hospital should be established and maintained. If the facility is located in the United States, depending on the type and amount of hazardous materials on-site, a representative may be required to interact with the local emergency planning committee (LEPC), as mandated by the EPA regulations, SARA Title III. State and federal agencies should also be integrated into the plan as appropriate (e.g., the U.S. Coast Guard).

Second, a facility may have to reach out to a community to share publicly available information on its operations and the associated storage and use of hazardous materials and their risks. The importance of this information is related to land use patterns and the potential for off-site risk. If a facility is remotely situated and the off-site hazards are minimal, then there is less need to inform the public. However, if a facility's property lines border directly on densely populated residential areas and there is significant off-site risk, then the need to share information is great. The facility must educate the public about the risks and associated emergency response procedures, including equipment and techniques for notification, as well as actions they should take to protect themselves.

6.5. Drills and Simulations

A carefully designed and well-written emergency response plan that sits on a shelf is useless. To be effective it must be rehearsed periodically. Rehearsal ensures that all involved personnel and organizations will respond in accordance with the plan and that any defects in the plan will be found and corrected.

Training and rehearsing can be accomplished in various stages, as described in the following sections.

6.5.1. Table-Top Exercises

A drill or simulation of a possible release is perhaps the most valuable training tool for identifying deficiencies in an emergency response plan. Drills can consist of a simple table-top exercise to realistically recreate an emergency that would involve off-site notification and support. Such an exercise is a good start in educating the plant staff on the intricacies involved in responding to an emergency. In conducting the exercise, staff are assigned different team roles to act out during the emergency and then the instructor introduces a viable emergency that could occur at their facility.

At first, the exercise is allowed to run its course with minimal input from the instructor. To make the exercise more realistic, however, the instructor occasionally updates the emergency by announcing an increase in its size or limitations on response resources, such as firewater or communications.

At the conclusion of the exercise, the team critiques its own performance and identifies deficiencies in the collective response to the emergency. To increase the stress level on the team, the instructor can update the emergency more frequently, increase its complexity, or complicate the situation by adding more factors, such as:

- The media
- Effects to the local populace
- Uncoordinated mutual aid
- Hazardous waste or asbestos exposure
- Loss of firewater supply
- Firewater/chemical runoff.

Table-top exercises are relatively inexpensive to perform and they produce a wealth of information on the capabilities and deficiencies of an emergency response plan. Deficiencies found during these exercises should be documented and the emergency response plan modified to correct them.

6.5.2. Plant-Wide Emergency Drills

The more realistic the emergency drill, the more valuable it will be in developing a good emergency response plan for postrelease mitigation. Table-top exercises can easily be expanded into more realistic simulations by recreating the emergency on a plant-wide scale. These drills should be conducted on off-shifts as well, when personnel resources are more limited. As in the table-top exercises, a credible emergency is envisioned, but in this case the actual emergency response systems are put into use, including:

- Communications
- Fire brigade and equipment
- Incident command centers
- Media rooms

- Evacuation and accountability procedures
- Equipment staging and plant security
- Off-site monitoring and notification.

Again, deficiencies identified during these exercises should be documented through a follow-up review meeting, and/or viewing of videos made during the exercise. The emergency response plan should then be modified to correct these deficiencies.

6.5.3. Full-Scale Emergency Simulations

Having identified any obvious deficiencies during table-top exercises and plant-wide drills, and having made the necessary corrections to the plan, plant management might then find it beneficial to conduct full-scale simulation exercises. They are a very effective means of evaluating the actual multidisciplinary and interrelated participation levels necessary to respond to an emergency. Involving those public authorities who would be expected to respond to an emergency can help to ensure an integrated response. Conducting such an exercise can be quite expensive, however, so it would be cost-effective to videotape the activities for future critiques and discussions. Following a critique, deficiencies in the emergency response plan should be corrected, and the emergency response team instructed in the revised plan using the sequence of exercises and drills just described.

6.6. Temporary Havens

A temporary haven can be a room or building that, because of its design and construction, can provide protection to a few people for the expected duration of a toxic release. Usually control rooms in plants are set up as safe havens for operators, to enable them to effect an orderly and safe shutdown of the facility and isolate the source of the release. These control rooms are usually fitted out with special HVAC systems that can be shut down quickly, air intakes that close tightly, and a separate supply of breathing air capable of supporting the necessary personnel for an extended period of time.

Another group of shelters can be classified as temporary havens. They do not provide the same level of protection as a safe haven but can provide shelter at the start of a crisis. A temporary haven is also a room or building that can provide protection from toxic fumes, but for a limited period of time. This may be adequate for toxic gas releases of short duration. However, means of escape from temporary havens must be provided, in case a hazardous situation is of long duration.

In emergency response plans, distinction between safe havens and temporary havens should be clear.

6.6.1. Criteria for Use

The major reason for considering the use of a safe haven or temporary haven is that immediate evacuation of the people affected may not be feasible. Gardner and Jann (1990) have shown that for chlorine and ammonia leak scenarios where the release duration is ten minutes, indoor sheltering can reduce the risk by 98.5% when compared to outdoor exposure. In this scenario, it was assumed that the building used as the temporary haven had an air change rate of once per hour.

Figures 6.7 and 6.8 are taken from Jann (1989) and were prepared to assess quickly the impact of various outside challenges on safe time by selecting the appropriate infiltration rate or air changes per hour (ACH) for the haven, and safe inside concentration. It must be noted that wind speed has not been accounted for in these curves. Higher wind speeds will cause high ACH for most buildings, thereby reducing the safe time for the haven.

Procedures and criteria for evacuation versus the use of temporary havens must be developed during the emergency response planning process. Such decisions should not be made casually at the time of the event.

The need for plant operation continuity will dictate whether the control room must be designed or modified to serve as a safe haven for the plant's operators. An evaluation must be made of the benefits to be gained by ensuring operating continuity, isolating that portion of the plant where the release is located, or shutting down the facility in an orderly manner. If operations are to continue, special protective equipment, breathing air supplies, and HVAC system operation must be provided for.

Concerning plant staff and the surrounding community, the use of temporary havens may be a viable option that will reduce the risk of toxic gas injuries. Gardner and Jann (1990) present techniques for quantitatively evaluating the effectiveness of temporary havens at specific community locations. These techniques involve first developing a set of release scenarios, followed by atmospheric dispersion modeling. The modeling must utilize site-specific meteorology. Lastly, the following information is needed: shelter-specific data, including building-specific infiltration rates (air changes per hour), and the toxicological properties of the chemicals involved (in the form of probit equations). With this information, a

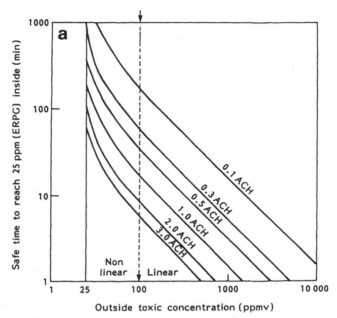

Figure 6.7. Temporary safe haven time to reach 25 ppm (ERPG) versus toxic challenge (Jann, 1989).

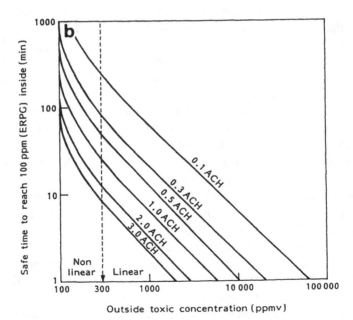

Figure 6.8. Temporary safe haven time to reach 100 ppm (ERPG) versus toxic challenge (Jann, 1989).

quantitative assessment may be made of the shelters available to the community and their effectiveness in the event of a toxic release. This assessment will determine whether the use of temporary havens for the community is a viable option.

6.6.2. Design Criteria

As stated above, the purpose of a temporary haven is to delay the infiltration of toxic gases into the area where people have gathered. Factors that influence the infiltration of toxic gases are the air changes per hour (ACH) that occur in the building, and the shutdown and startup of the heating, ventilation and air conditioning (HVAC) systems of the building.

The ACH is normally a function of a building's HVAC system and the amount of fresh air it brings into the building each hour divided by the volume of the building. Since the HVAC system will be shut down during a toxic gas incident, the ACH during this time is a measure of the rate of toxic gas infiltration into the building. This in turn indicates the toxic exposure of the people in the shelter. Sulfur hexafluoride, SF_6, has been used as a tracer to determine the air infiltration rates into buildings (Jann, 1988, 1989). In this approach the tracer is introduced into the building's HVAC system and allowed to mix thoroughly. After an adequate time for mixing, the HVAC system is shut down and gas samples are taken to measure the concentration of the tracer SF_6. The decline in the concentration of SF_6, over time is a measure of the infiltration of outside air into the building.

Jann (1989) reports values of ACH measured for industrial buildings and office buildings ranging from a low of 0.10 to a high of 1.09 for a building in which lab hoods and computer room air conditioners were not shut down during the test. He found that, generally, older buildings had higher ACH values (1.3 to 3.1) than newer constructions.

Typical infiltration values for housing in North America vary by a factor of about ten. For tight housing, the values range from 0.2 air changes per hour to a high of 2.0 (ASHRAE, 1985). Two surveys of average infiltration rates of North American housing have been done. In the first, 312 homes were evaluated and a median infiltration rate of 0.5 ACH was found. This group consisted mainly of new energy-efficient houses. The second survey consisted of 266 houses, with a 0.9 ACH median value. This group represented older, low income housing (ASHRAE, 1985, pp. 22-7–22-17).

Remedial measures that will reduce the infiltration rates into building are listed in Table 6.2.

TABLE 6.2
Infiltration Remedial Checklist (Jann, 1989)

Building Walls   *Seal around all through-the-wall penetration and weather hoods for:*   • Ducts   • Conduits   • Electrical/instrument wiring trays   • Pipes   *Reduce infiltration through construction of*   • Masonry walls. Mortar joints, lintels and through-the-wall flashing must be tight.   • Sheet or panel (metal, plastic, etc.) wall siding. Seal tight all lap or butt joints. Make certain that caps, covers and flashings don't have hidden voids due to siding configurations (flutes, corrugations, etc.)   *Protect from infiltration through open trenches:*   • Pipe or conduit   • Waste water   *Wall louvres:*   • Make sure that blades can close tight.   • Make sure that frame is sealed.   • Determine whether automatic emergency damper may be required.
Exterior doors   *Reduce infiltration at personnel doors by*   • Installing weather strip seals including automatic door bottom   • Using automatic door closers without hold-open devices   • Each time a door is opened, about 1000 ft^3 of air can infiltrate. Vestibule door system can make 40% reduction. Vestibule depth must ensure that first door closes before second door is opened   *Reduce infiltration at large access doors such as overhead or sliding type by*   • Installing jamb, head and bottom weather strip seals   • Keeping doors normally closed
Windows   *Windows should be constructed so glazing and frame joints are gasketed or sealed against infiltration. Keep in mind:*   • Fixed nonoperating windows are preferred.   • Some international building codes may require operating windows for ventilation. Consider automatic release overhead shutters for emergency closing of window openings.   • The preferred design includes no windows or a minimum number of windows.
Roofs   • Eave closures, cants, flashings and facias between roof and wall must not have voids that can allow toxic gas/vapor infiltration.   • Through-the-roof ducts and pipes must have space between duct or pipe and curb adequately sealed.   • Platform or equipment support flashing must be sealed so there can be no infiltration around and behind the flashing and into the building.   • Roof penetrations should be minimized to reduce potential for gas or vapor penetration.

At the same time that efforts are being made to reduce the infiltration into a building, effectiveness of the HVAC system must be evaluated to ensure that sick building syndrome will not be the result.

Studies have been made of the effect that the shutdown and startup of building HVAC systems or the opening up of a building can have on the total toxic load the people in the shelter are exposed to (Chikhliwala et al., 1989, 1993). These analyses show that a delay in the shutdown of the HVAC system, as well as the restart of the system after the toxic cloud has passed, can result in an increase in the toxic load. The examples evaluated show that sheltering in place can reduce indoor concentrations by a factor of 2 to 10 when compared to outdoor values. This indicates that prompt shutdown of HVAC systems and air intakes can substantially reduce maximum concentrations and dosage in the haven, while a delay in restarting the system can contribute to a higher dosage than necessary.

6.6.3. Capacity

The number of people to be located in each safe haven or temporary safe haven must be carefully evaluated and planned for. This is to ensure that there will be an adequate number of escape air masks for those taking shelter in the buildings if the incident lasts longer than anticipated over time. In a situation like this, the toxic concentration will rise over time and vacating the haven will be necessary. When planning for the number of escape masks needed, some allowance must be made for visitors who may be seeking shelter as well.

Concerning use of residential or commercial buildings around a plant site for protection of neighboring populations, the capacity and the likely number of people in each residence or building should be known. Again, should the release last longer than anticipated, plans and vehicles must be in place to evacuate these people quickly.

6.6.4. Communications and Other Equipment

It is recognized that effective communications are vital to an effective and timely response to an emergency situation. This principle also applies to people located in safe havens and temporary safe havens. In cases where a plant control room has been set up as a safe haven for the operators, they must be able to communicate by radio with response personnel throughout the plant site. If the operators are to wear breathing apparatus then the radios must fit inside the breathing air masks.

Plant personnel in temporary safe havens must be able to communicate by radio with the response control center. The radio connection will facilitate communication of information on the number of people in the shelter and any medical needs that they might have. It will also enable the control center to notify the occupants when it is clear to restart HVAC systems and leave the shelter.

In addition to escape masks, shelters should contain air monitoring equipment to allow the determination of the level of toxic gas in the shelter atmosphere (see Section 6.2.3). Also, provision of emergency lighting and first aid supplies and equipment should be made.

6.7. References

29 C.F.R. sec. 1910.38. Occupational Safety and Health Administration.

29 C.F.R. sec. 1910.146. Occupational Safety and Health Administration.

29 C.F.R. sec. 1910.120. Occupational Safety and Health Administration.

29 C.F.R. sec. 1910.119. Occupational Safety and Health Administration.

ASHRAE (American Society of Heating, Refrigeration and Air-conditioning Engineers) 1985. *ASHRAE Handbook 1985—Fundamentals.* Atlanta: American Society of Heating and Air-Conditioning Engineers, Inc.

CCPS (Center for Chemical Process Safety). 1988. *Guidelines for Safe Storage and Handling of High Toxic Hazardous Materials.* New York: American Institute of Chemical Engineers.

CCPS (Center for Chemical Process Safety). 1989. *Guidelines for Technical Management of Chemical Process Safety.* New York: American Institute of Chemical Engineers.

CCPS (Center for Chemical Process Safety). 1995. *Guidelines for Technical Planning for On-Site Emergencies.* New York: American Institute of Chemical Engineers.

CFR, Title 68, 1993. U.S. Environmental Protection Agency. (January 18).

Chikhliwala E. D., P. R. Jann, and S. Kothandaraman. 1993. Assessing the Effectiveness of Safe Havens for Toxic Chemical Releases. Paper presented at the Summer Meeting of the American Institute of Chemical Engineers, Seattle WA.

Chikhliwala E. D., M. Oliver, and P. R. Jann. 1989. Exposure Mitigation Shelters during a Toxic Gas Release. In L. J. Brasser and W. C. Mulder. (Eds.) *Man and His Ecosystem: Proceedings of the 8th World Clean Air Congress at The Hague, Netherlands, September, 11–15,* Vol. 1, pp. 265–270. Amsterdam: Elsevier Science Publishers.

Gardner R. J. and P. R. Jann, 1990. Reduction of Community Risk by Indoor Sheltering Qualitative Evaluation and Application. Paper presented at the 83rd Annual AWMA (American Waste Management Association) Meeting, Pittsburgh, PA.

Jann P.R. 1988. Evaluation of Temporary Safe Havens. Paper presented at the 81st Annual Meeting of the Air Pollution Control Association, June 19–24, Dallas, TX.

Jann P. R. 1989. Evaluation of Temporary Safe Havens. *Journal of Loss Prevention Process Industries.* 2 (January): 33-38.

7

Examples of Mitigation Effectiveness

7.1. Introduction

Consequence modeling can be used to evaluate the impact of post-release mitigation measures and determine the relative effectiveness of techniques, or combinations of techniques. Release scenarios are described in Chapter 2, and mitigation techniques are presented in Chapters 3 through 6.

This chapter focuses on four examples that have been selected to demonstrate the effects of mitigation measures. These examples are for demonstration purposes only; they do not necessarily represent the optimum design of a mitigation system or actual engineering practices. The four mitigation methods that will be described are:

- the use of a dike to contain a spill,
- the use of foam to cover a spill,
- the use of refrigeration to reduce the effects of a release, and
- the use of water sprays to reduce the effects of a release.

7.2. Consequence Modeling

Consequence modeling, for the purposes of the illustrations given in this chapter, means the prediction of ambient atmospheric concentrations using models for quantifying the release of fluids from containment, and the formation of vapor and liquid aerosol plumes using dispersion models.

First we will look at a release scenario that is unmitigated, and then at the modification of a scenario to include a postrelease mitigation technique. The effect of the postrelease mitigation technique will be evaluated by applying the consequence modeling techniques described above. It is important to note that consequence modeling techniques are based on the premise that correct mathematical representation of all-important processes, such as conservation of mass, energy, momentum, the system components and changing physical properties with changing temperatures, is possible and appropriate. Thus, a good understanding of the models being used, and their accuracy relative to the scenarios being evaluated, is critical.

The determination of the consequences for a group of selected release scenarios may be found in the CCPS document entitled *Guidelines for Use of Vapor Cloud Dispersion Models, Second Edition* (CCPS, 1996).

7.3. Basis for Examples

The examples that are used in the following sections, and the conditions that have been selected to illustrate the effectiveness of the postrelease mitigation techniques, may not represent standard industrial practices for the materials selected. For detailed engineering information the manufacturers or suppliers of the materials should be consulted. They can also supply information on the most effective postrelease mitigation techniques to use in the event of an accidental loss of containment.

Also bear in mind that the meteorological and other information described in the following section has been selected as representative conditions only. When an actual situation is being considered the meteorological conditions unique to the site under consideration should be used.

For each of the postrelease mitigation techniques examined a description of the consequences of the unmitigated release will be presented first, followed by the mitigated release. The results have been tabulated and shown in figures.

7.4. Modeling Conditions

For each of the examples, two sets of meteorological conditions will be used. They will be applied to both the unmitigated and mitigated case. The tanks in the examples have been located in the vicinity of Houston, Texas in an open area. The meteorological conditions listed in Table 7.1 are considered typical of what might be encountered during a daytime and nighttime release. A Pasquill D-stability (neutral) with a wind speed of 5 m/s (11 mph), or [D/5], is typical of daytime conditions. During nighttime hours, the atmosphere is more stable and lower wind speeds are found,

TABLE 7.1
Modeling Meteorological Conditions

Condition	Set 1	Set 2
Pasquill Stability	D	F
Wind Speed	5 m/s	2 m/s
Air Temperature	25°C	15°C
Humidity	70%	90%
Terrain Roughness	0.1 m	0.1 m
Time of day	1400 hr	0230 hr
Sky	clear	clear

making the Pasquill F and wind speed of 2 m/s (4.5 mph) [F/2] representative (CCPS, 1989). When a specific site is being modeled, meteorological data can be secured from the National Oceanographic and Atmospheric Administration (NOAA), National Climatic Data Center, or from a nearby airport.

For illustrative purposes, the hazardous properties of the materials used in the examples have been modeled to the Emergency Response Planning Guideline, ERPG-2 (see Section 3.1.1) concentration for toxic materials, or to one half of the lower flammable limit (LFL/2) for flammable material. Lastly, only the first 3600 seconds of the event are considered in these examples. For an actual situation, different concentration levels for toxic or flammable concentrations may be considered and different durations as well.

7.5. Effect of Diking

To demonstrate the effectiveness of diking as a postrelease mitigation measure, an accidental release of carbon disulfide from a vertical storage tank will be evaluated. In the example, isopleths to the ERPG-2 concentration of 50 ppm (AIHA, 1992) will be considered.

For the mitigated and unmitigated cases the carbon disulfide is contained in a vertical steel tank having a diameter of 4.3 m and a height of 6.5 m. The accidental release comes from a 50-mm diameter nozzle located at the base of the tank. The tank is filled to 60% of capacity. The tank has a low-pressure nitrogen purge.

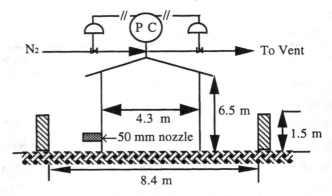

Figure 7.1. Storage tank with dike (carbon disulfide and *n*-pentane).

On the basis of the criteria discussed in Section 5.2 and engineering judgment, a square dike, having an 8.4-m side and height of 1.5 m was selected to show its effect on the accidental release. See Figure 7.1.

The first hour of the incident will be modeled. This is consistent with the ERPG exposure times. See Section 3.1.1

In both cases the discharge flow rate from the storage tank is the same. As the height of liquid in the tank decreases the available head on the outlet decreases and, therefore, the flow rate also decreases. This effect is shown in Figure 7.2.

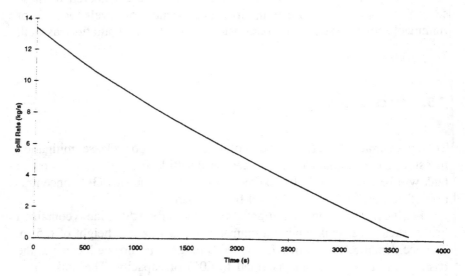

Figure 7.2. Carbon disulfide release rate from storage tank.

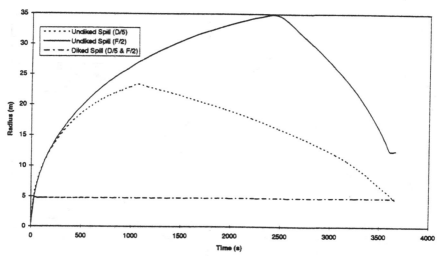

Figure 7.3. Carbon disulfide pool radius for the unconfined and diked scenarios.

As the carbon disulfide discharges from the failed nozzle at the base of the tank, it will form a pool from which the material will vaporize. Figure 7.3 shows the radii of the resultant pool for the unmitigated case (no dike) at the two meteorological conditions being evaluated, Pasquill D at a 5 m/s wind velocity (D/5), and Pasquill F at a 2 m/s wind velocity (F/2). As this figure shows, the pool resulting from the spill at the D/5 conditions reached a maximum diameter 1000 seconds into the incident while the spill at the F/2 condition reached its maximum size 2500 seconds into the incident. For the same situation involving the dike under both conditions the dike floor was covered almost immediately after the start of the release, 30 seconds, and the area of the spill surface remained constant as would be expected.

The reason for the differences in the pools formed by the spill in the unmitigated case can be explained by the curves for the evaporation rates of carbon disulfide. See Figure 7.4. First, consider the unmitigated D/5 case. In this case the maximum evaporation rate from the pool is reached approximately 1000 seconds after the start of the incident and is 8.5 kg/s, the largest of any of those considered. Factors that influenced this vaporization rate are listed below.

- The pool, as shown in Figure 7.3, has reached its maximum diameter. Therefore, the maximum surface area for evaporation is present.
- Correspondingly, after this the rate of material input into the pool is dropping below this evaporation rate. At approximately 1200 seconds, as shown by Figure 7.2, the release rate drops below the

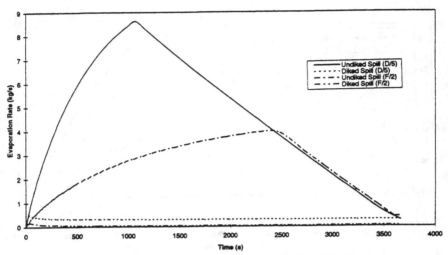

Figure 7.4. Carbon disulfide evaporation rates for undiked and diked scenarios.

evaporation rate of 8.5 kg/sec, causing the pool to start to shrink as it is losing material faster than it is being added. The result is a reduction in surface area for vaporization.

- However, the most critical factors for the high evaporation rate are the meteorological conditions. A high, 5 m/s wind velocity over the surface of the spill, promotes mass transfer into the air. Also, this spill has happened in the daytime, (14:00 hr), onto a hot concrete surface, which results in large heat input that provides the needed energy for vaporization, as explained in Section 3.1.2.1. Lastly, the pool is receiving energy input from the sun.

As can be seen in Figure 7.4, the situation for the unmitigated F/2 conditions is very different from the D/5 situation. As Figure 7.3 shows, a much larger pool is formed for the F/2 while Figure 7.4 shows that much lower evaporation rates from the pool occur later in the incident at 2500 seconds. Important factors affecting this event are listed below.

- Not until 2700 seconds into the event has the discharge rate from the tank decreased to less than the evaporation rate. This means that until this time the pool is continuing to grow, providing the maximum surface area for evaporation.
- Owing to the lower wind velocity (2 m/s versus 5 m/s), there is a lower rate of mass transfer and heat transfer from the atmosphere to the pool surface.

- Because this event is occurring at night, even though onto the same surface, there is less heat input from the surroundings to promote vaporization. As was noted in Section 3.1.2.1, if steps are taken to shade surfaces onto which a spill may occur, or shade the surface of the resultant pool, the rate of vaporization can be reduced. These two scenarios illustrate the effectiveness of shading.

In the mitigated case, in which the spilled carbon disulfide is retained by the dike installed around the tank, there is a large reduction in the vaporization rates of carbon disulfide for both sets of meteorological conditions. The primary reason for the differences is that the available surface area for vaporization is significantly reduced. Second, the vaporization rate remains essentially constant for the duration of the incident because the surface area for vaporization is constant. The small variations that do show in these curves are due to small effects of heat transfer into the diked carbon disulfide, as discussed in Chapter 5.

The hazard zones for each of the cases considered were calculated using a dispersion model. The one selected was the SLAB model, which is available in the public domain. The input into the dispersion model was the rate of evaporation of the carbon disulfide from the pool, as shown in Figure 7.4, and the meteorological conditions given in Table 7.1. The models then calculated the isopleths to the desired concentration level, which in this case is the ERPG-2 value of 50 ppm. Results of these calculations are tabulated for the two unmitigated and the two mitigated cases in Table 7.2. The isopleths are shown graphically in Figures 7.5 to 7.8.

The results in Table 7.2 and Figures 7.5 to 7.8 show that for the D/5 conditions, even though there is a higher peak vaporization rate, the hazard area is much reduced when compared to the F/2 conditions. The major reason for this is that the higher turbulence and wind speed of the D/5 conditions result in more rapid mixing and dilution of the carbon disulfide vapors in the atmosphere. Under the more stable F/2 conditions mixing does not occur rapidly, resulting in larger hazard zones. As Figures 7.5 to 7.8 and the results in Table 7.2 show, a dike around a tank can be a useful postrelease mitigation measure that has the potential to greatly reduce the impact of such an incident.

Further mitigation and reduction of the hazard zones might be obtained by incorporating some of the other dike enhancing features discussed in Section 5.2.1.

TABLE 7.2
Carbon Disulfide Example

Mitigation by Diking: ERPG-2 Hazard Zones		
Unmitigated Scenario		
Wind Speed, m/s	5.0	2.0
Pasquill Stability	D	F
Temperature, °C	25	15
Relative Humidity, %	70	90
ERPG-2 Hazard Zones		
Zone Length, m	750	3200
Zone Width, m	140	480
Zone Area, m^2	80,000	1,230,000
Mitigation by Diking		
Wind Speed, m/s	5.0	2.0
Pasquill Stability	D	F
Temperature, °C	25	15
Relative Humidity, %	70	90
ERPG-2 Hazard Zones		
Zone Length, m	150	230
Zone Width, m	30	35
Zone Area, m^2	3500	6300

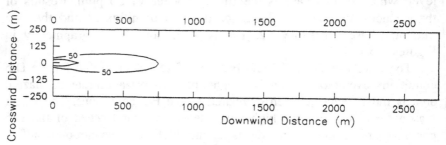

Figure 7.5. Hazard zone for ERPG-2 (50 ppm) for undiked carbon disulfide release D/5.

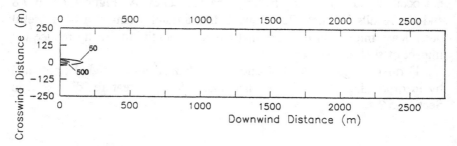

Figure 7.6. Hazard zone for ERPG-2 (50 ppm) for diked carbon disulfide release D/5.

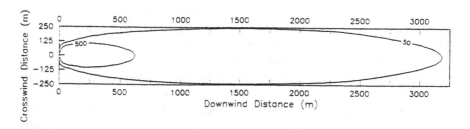

Figure 7.7. Hazard zone for ERPG-2 (50 ppm) for undiked carbon disulfide release F/2.

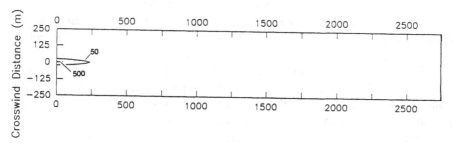

Figure 7.8. Hazard zone for ERPG-2 (50 ppm) for diked carbon disulfide release F/2.

7.6. Use of Foam

In this example the use of foam, which was described in Section 3.3.1, will
be demonstrated as a postrelease mitigation agent. An accidental release of
n-pentane, a flammable material, from a storage tank into a diked area, and
a hazard zone to a concentration of one half the lower flammable limit
(LFL/2) will be evaluated. For *n*-pentane the LFL is 1.4% by volume in air.
Then the LFL/2 will be 0.7% by volume in air (GPSA, 1972).

The *n*-pentane in this example is stored in a vertical steel tank having
a diameter of 4.3 m and a height of 6.5 m and is shown in Figure 7.1. The
accidental release occurs from a 50-mm diameter nozzle at the base of the
tank. When this accident happened the tank is 60% full. The discharged
n-pentane for the unmitigated case is contained within a square dike with
8.5-m sides and 1.5 m high. Storage in the tank is under a minimal pressure
imposed by the nitrogen blanketing system.

In the mitigated case, foam is applied to the surface of the spill 15
minutes after the release of the *n*-pentane has started. In this example it has
been assumed that a 50% reduction in the vaporization rate off the surface

of the *n*-pentane will be achieved by the application of foam. For actual reductions of the vaporization rate foam suppliers should be consulted as discussed in Section 3.3.1 (Vapor Suppression Foams).

The meteorological conditions listed in Table 7.1 apply to both cases. For the calculation of vapor dispersion from the diked pool the SLAB dispersion model was used.

For both releases the growth of the pool is shown in Figure 7.9. The figure shows that the surface inside the dike is quickly covered after the release of *n*-pentane and the maximum surface area for vaporization remains constant for the full 3600 seconds of the incident.

Figure 7.10 shows the evaporation rate from the pool inside the dike for the mitigated and unmitigated releases at D/5 conditions. As can be seen from this figure, the initial phase of the incident is the same for both situations. Once the foam is applied to the surface of the pool (900 seconds after the incident has started) there is a reduction in the rate of *n*-pentane vaporization. The figure shows that the vaporization rate will remain constant for the duration of the incident. This assumption is valid, as was pointed out in Section 3.3.1, as long as new foam is applied to the pool to keep the foam blanket in place. Figure 7.11 shows a similar situation for the F/2 conditions scenarios except the vaporization rates are lower.

Calculation of the hazard zone for the *n*-pentane vaporized from the spill, for both scenarios, is summarized in Table 7.3. The information in Table 7.3 indicates that at the D/5 conditions the hazard zone for both the

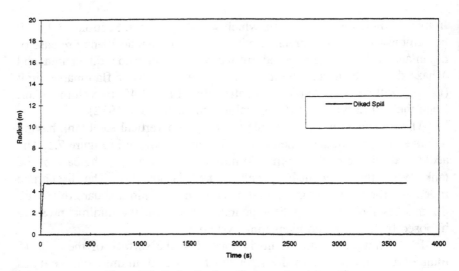

Figure 7.9. Pool radius for spill of *n*-pentane into a dike.

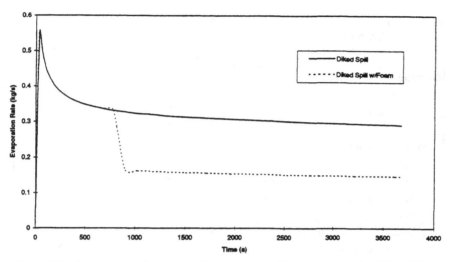

Figure 7.10. Evaporation rate for n-pentane for uncovered and foam covered pool at D/5 conditions.

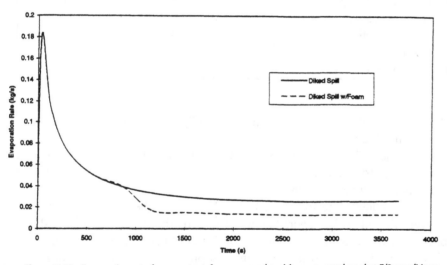

Figure 7.11. Evaporation rate for n-pentane for uncovered and foam covered pool at F/2 conditions.

unmitigated and mitigated situations is located over the surface of the n-pentane pool contained inside the dike. This can be attributed to the higher wind speed and turbulence of these conditions which result in the rapid dilution of the n-pentane vapors coming off the surface of the pool. Despite rapid dilution, it is clear that there are vapors in the flammable range above

TABLE 7.3
n-Pentane Example

Mitigation by Foam Application		
Unmitigated Scenario		
Wind Speed, m/s	5.0	2.0
Pasquill Stability	D	F
Temperature, °C	25	15
Relative Humidity, %	70	90
Hazard Zone to ½ LFL		
Zone Length from edge of dike, m	0.0	13
Zone Width from edge of dike, m	0.0	15
Zone Area from edge of dike, m^2	0	150
Mitigation by Foam Application		
Wind Speed, m/s	5.0	2.0
Pasquill Stability	D	F
Temperature, °C	25	15
Relative Humidity, %	70	90
ERPG-2 Hazard Zones		
Zone Length, m	0.0	0.0
Zone Width, m	0.0	0.0
Zone Area, m^2	0	0

the pool (inside the dike) and that a small spark could still result in a hazardous situation.

For the F/2 conditions, the results in Table 7.3 show that there is a hazard zone for the unmitigated situation (or until the foam is applied) that extends downwind from the tank.

This is due to the stability and low velocity of the air, which results in slower dilution of the vapors below the LFL by the air. As the results show, once the foam is applied the hazard zone is confined to the diked area. This also implies that the earlier the foam is applied, the sooner the hazard zone can be reduced and the hazard controlled.

7.7. Mitigation by Refrigeration

As was discussed in Section 3.2, if there are process reasons for refrigeration, it can also be effective in reducing the consequences of a release. In this example the release of ammonia from a vertical storage vessel has been chosen for the example. The unmitigated case will be the accidental release of anhydrous ammonia being stored under it own vapor pressure. In the mitigated case which features refrigeration, the anhydrous ammonia is cooled to –33°C. After demonstrating the effect refrigeration has on mitigating the release, the use of an additional postrelease mitigation measure, diking, in conjunction with the refrigerated storage will be evaluated.

For both the pressurized and refrigerated storage scenarios, the vessel consists of a vertical cylindrical tank 1.22 m in diameter and 4.88 m high. The tank is mounted on legs which raise the bottom tangent line 1.53 m above the grade. The scenarios involve the accidental striking and shearing of a bottom outlet pipe that is 37.5 mm in diameter. Upstream of the break point the line contains two valves. At the time of the incident the tank is 60% full. The tank is shown in Figure 7.12.

7.7.1 Pressure Storage of Ammonia

The liquid ammonia is stored in the tank at a temperature of 20°C under its vapor pressure of 851 kPa. Under these conditions the discharge calculations show that all of the ammonia in the vessel is released to the

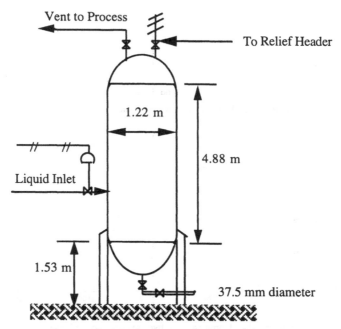

Figure 7.12. Anhydrous ammonia storage tank system.

atmosphere from the broken pipe and is immediately vaporized. The total time of the release for the tank contents is 375 seconds.

The SLAB dispersion model was used for this situation. Figures 7.13 and 7.14 show isopleths for the two sets of meteorological conditions (D/5 and F/2) for a 3600-second averaging time. As in the carbon disulfide and n-pentane example, the more stable meteorological conditions (F/2) result in the larger hazard zone.

7.7.2. Refrigerated/Ammonia Storage

For this example the anhydrous ammonia is stored as refrigerated liquid at a temperature of $-33°C$ and one atmosphere of pressure. The tank is vented to a system which recovers any vapors that boil off. The configuration of the tank is identical to that shown in Figure 7.12, and the accident causing the release is the same as in the pressurized storage example described above.

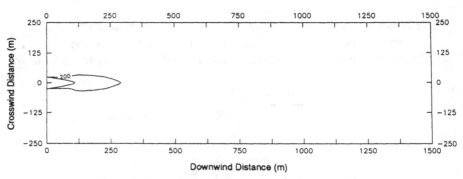

Figure 7.13. Unmitigated anhydrous ammonia release D/5.

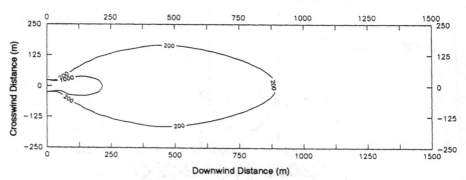

Figure 7.14. Unmitigated anhydrous ammonia release F/2.

When the 37.5-mm diameter pipe fails, the liquid ammonia is released from the tank and forms an unconfined pool on the ground. Figure 7.15 shows the growth of the pool for both the D/5 and F/2 meteorological conditions. As was shown in the carbon disulfide example, the pool formed at the D/5 conditions is smaller than the one for the F/2 conditions. The reasons in this case are the same as those discussed in the carbon disulfide example (see Section 7.5). Vaporization rate as a function of time is given in Figure 7.16.

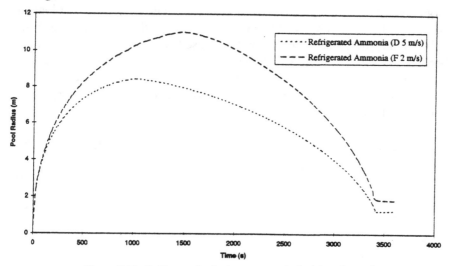

Figure 7.15. Refrigerated anhydrous ammonia liquid pool growth.

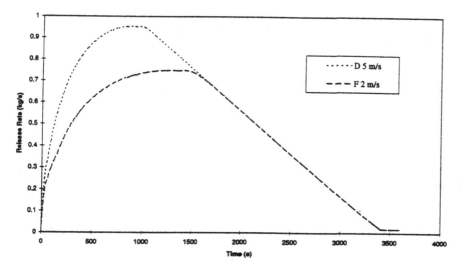

Figure 7.16. Vaporization rate from undiluted pool as function of time.

The dispersion modeling of the resultant ammonia vapor hazard zones to the ERPG-2 value for ammonia of 200 ppm was done using the SLAB model. The isopleths for the D/5 and F/2 meteorological conditions found in these examples are shown in Figure 7.17 and 7.18.

The results for the warm and pressurized ammonia storage (unmitigated) and the refrigerated ammonia storage (mitigated) are summarized in Table 7.4.

A comparison of the hazard zones for both storage conditions suggests that the use of refrigeration as part of the mitigation system did not yield a large benefit. The area of the hazard zones for both sets of meteorological conditions have been reduced by only 25 to 30%, and distances to ERPG-2 only slightly decreased for D/5 and increased for F/2 conditions. This is because in the unmitigated situation all of the ammonia was rapidly released and formed a large cloud. In the refrigerated case the ammonia was released over the 3600 seconds of the incident and formed a pool from which it was continuously vaporized.

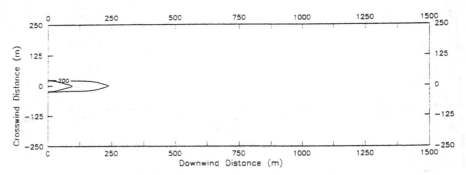

Figure 7.17. Hazard zone from refrigerated anhydrous ammonia release for ERPG-2 concentration at D/5 conditions.

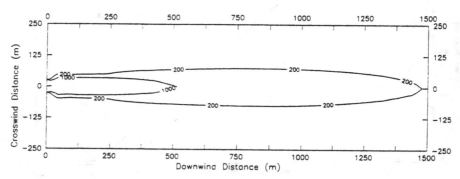

Figure 7.18. Hazard zone from refrigerated anhydrous ammonia release for ERPG-2 concentration at F/2 conditions.

TABLE 7.4
Anhydrous Ammonia Example

Mitigation by Refrigeration		
Unmitigated Scenario: Pressurized Ammonia Storage		
Wind Speed, m/s	5.0	2.0
Pasquill Stability	D	F
Temperature, °C	25	15
Relative Humidity, %	70	90
Zone Length, m	290	890
Zone Width, m	60	330
Zone Area, m^2	12,000	200,000
Refrigerated Ammonia Storage		
Wind Speed, m/s	5.0	2.0
Pasquill Stability	D	F
Temperature, °C	25	15
Relative Humidity, %	70	90
Zone Length, m	230	1500
Zone Width, m	45	140
Zone Area, m^2	8000	156,000

7.7.3 Refrigeration Combined with Diking

As discussed in Section 3.1.2.1, a liquid that is uncontained is one over which there is no control and which will result in potentially severe consequences. If a dike is placed around the tank containing the refrigerated liquid ammonia and the ammonia spill is confined within it, a much reduced hazard zone can be obtained because we have limited the surface area available for vaporization and additional postrelease mitigation measures can be applied. As pointed out in Chapter 3, combinations of postrelease mitigation measures will provide the best overall response to an accidental release.

In this scenario, a dike 8.4 m by 8.4 m having a height of 1.5 m is placed around the refrigerated ammonia storage tank. The accidental failure of the 37.5-mm line occurs inside the diked area. All of the liquid ammonia released from the tank is contained inside the dike. The rate of flow of the liquid ammonia into the dike is the same as for the refrigerated example (Section 7.7.2).

The isopleths for the resulting hazard zone to the ERPG-2, 200 ppm concentration are shown in Figures 7.19 and 7.20. A comparison to Figures 7.13 and 7.14, shows that the use of refrigeration plus a dike to collect the liquid ammonia reduces the surface area available for vaporization and the energy input for vaporization, which has a significant impact on the quantity of material getting into the atmosphere and the resulting hazard zones.

A comparison of the hazard zone areas in Table 7.5 shows that the combination of refrigerated storage with a dike results in significant reductions in the range of 50 to 85%. This example illustrates the point made in Section 3.1.2.3 that the most effective postrelease mitigation results will be obtained through the combination of the separate methods discussed in this guideline.

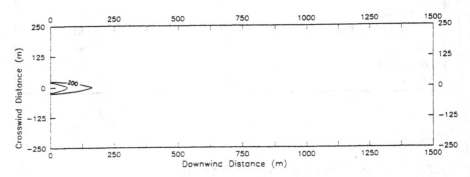

Figure 7.19. Hazard zone for refrigerated/diked anhydrous ammonia release for ERPG-2 concentration at D/5 conditions.

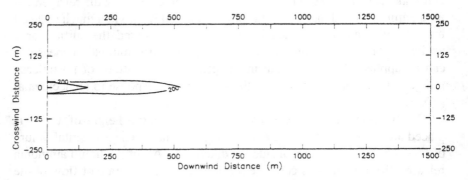

Figure 7.20. Hazard zone for refrigerated/diked anhydrous ammonia release for ERPG-2 concentration at F/5 conditions.

TABLE 7.5
Anhydrous Ammonia Example

Mitigation by Refrigeration and Diking		
Mitigated by Refrigeration and Diking		
Wind Speed, m/s	5.0	2.0
Pasquill Stability	D	F
Temperature, °C	25	15
Relative Humidity, %	70	90
Zone Length, m	160	520
Zone Width, m	30	50
Zone Area, m^2	3800	21,000

7.8. Use of Water Sprays

In this section we will demonstrate the use of water sprays as a postrelease mitigation measure. The theory and design of water spray systems has been detailed in Chapter 4. The scenario will involve the accidental release of hydrofluoric acid (HF) from a storage tank. Again we will be considering the distance to reach the ERPG-2 concentration for HF of 20 ppm (AIHA, 1992).

For both the mitigated and unmitigated case the HF will be released from a storage tank having a capacity of 29 m^3 (7600 gallons). The size of the hole in the storage tank is 19.1 mm (0.75 inches) due to the rupture of a pipe on the side of the vessel. The HF in the tank contains 2% water, on a weight basis, and is stored under a pressure of 1.4 atm abs. and temperature of 15°C. The ambient air temperature is 15°C with a 50% relative humidity. To illustrate the effectiveness of the water sprays used to mitigate this release the worst case meteorological conditions, a wind velocity of 1 m/s and a Pasquill F stability, will be considered.

Dispersion modeling of the unmitigated and mitigated scenarios was done using the HGSYSTEM (Version 2.4) (Witlox et al., 1991; Post, 1994). This modeling system was originally developed by the ad hoc Industrial Committee on HF Mitigation and Assessment with further work being done by the American Petroleum Institute. HGSYSTEM is a set of general models that have been modified to deal with the unique chemistry of HF.

Since the HF is being stored below its normal boiling point, it was assumed that no aerosols would be formed. All of the released HF will form a pool on the ground. The HF vapors that get downwind will come from evaporation from this pool in both the unmitigated and mitigated case. For the F/1 meteorological conditions the peak evaporation rate was found to be 0.93 kg/s while the average evaporation a rate for a 3600 second averaging time was found to be 0.77 kg/s.

The dispersion modeling for the unmitigated case predicts a distance of 5 km for the concentration to be reduced to 20 ppm, under F/1 weather conditions. As in the other examples the higher wind speed and turbulence of more severe meteorological conditions result in a more rapid dilution and a smaller hazard zone.

The water-spray system that has been installed for the mitigation of any HF releases that might occur from the vessel is a downward-facing spray pattern and based on work done by Schatz and Koopman (1989), Fthenakis et al. (1990), Blewitt et al. (1991), Petersen and Blewitt (1992), Fthenakis and Blewitt (1992), and Fthenakis and Blewitt (1994). For this water-spray curtain it was assumed that the large-scale field data collected during the Hawk experiments (Schatz and Koopman, 1989) provided the basis for assuming a removal efficiency of 86% of the HF when a water ratio of 40:1 on a volume basis is provided by the spray curtain. See Figure 4.2 in Chapter 4. This removal efficiency was used for the quantification of the benefits to be realized by using water sprays as a postrelease mitigation measure for a hydrofluoric acid release.

The difference in the amount of material going downstream for the mitigated and unmitigated cases is shown in Figure 7.21.

Figure 7.22 shows the effect the reduced input for the mitigated case, caused by the capture of HF by the water spray, has on the centerline concentration of HF to the ERPG-2 value of 20 ppm as compare to the unmitigated case to the same concentration.

In the analysis it was assumed that the water-spray curtain was activated within one minute of the start of the incident and that 86% of the HF vapor from the evaporating pool that reached it was removed. The balance of the HF not removed by the water sprays became the material that formed the reduced hazard zone and the input for the dispersion modeling. For the F/1 meteorological conditions the mitigated hazard zone for the centerline concentration of 20 ppm was 750 m. The effectiveness of water spray as postrelease mitigation measure is shown in Table 7.6.

As demonstrated by this example, the use of water-spray curtains can be effective in reducing the size of the hazard zone downwind of the release. For this situation to be modeled effectively, it is essential to know, as in the case of HF, the efficiency that can be expected for various water flow rates in relation to the quantity of material released.

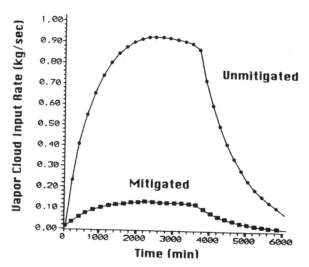

Figure 7.21. HF input rate to the dispersion model for the mitigated (water-spray curtain) and unmitigated scenario.

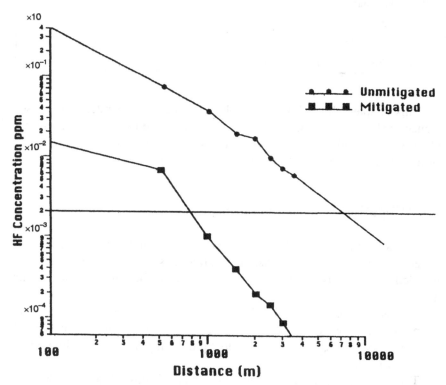

Figure 7.22. Plume centerline concentrations for a mitigated (water-spray curtain) and unmitigated HF spill scenario.

TABLE 7.6
Hydrofluoric Acid Example

Mitigation by Diking: ERPG-2 Hazard Zones	
Unmitigated Scenario	
Wind Speed, m/s	1.0
Pasquill Stability	F
Temperature, °C	15
Relative Humidity, %	50
Pool Evaporation Rates	
Peak, kg/s	0.93
3600-sec Average, kg/s	0.77
Haxard Zone Centerline Distance, km	5.0
Mitigation by Use of Water Sprays	
Wind Speed, m/s	1.0
Pasquill Stability	F
Temperature, °C	15
Relative Humidity, %	50
Haxard Zone Centerline Distance, m	0.75

This chapter, through the examples it contains, illustrates that consequence analysis done with the proper models can be useful in determining which of the available postrelease mitigation techniques will be effective for a specific scenario and provide the greatest benefits.

7.9. Mitigation System Selection

As discussed in Chapter 2, mitigation systems can be divided into two categories: pre- and postrelease mitigation systems. Both systems are intended to reduce risk by either reducing the probability an event will occur or the consequence if it does happen.

It must be recognized that these systems are the last line of defense against a hazardous event, and to be effective they must perform when called upon. In other words, there must be a low probability of system failure when a demand is placed upon it.

The primary objective of mitigation systems is cost-effective reduction of risk. A recommended systematic risk-based approach for the selection of mitigation systems is presented in Figure 7.23. The cycle begins by qualitative evaluation of pathways to hazards. This is typically achieved using hazard and operability studies (HAZOP), failure modes and effects analysis (FMEA), etc.

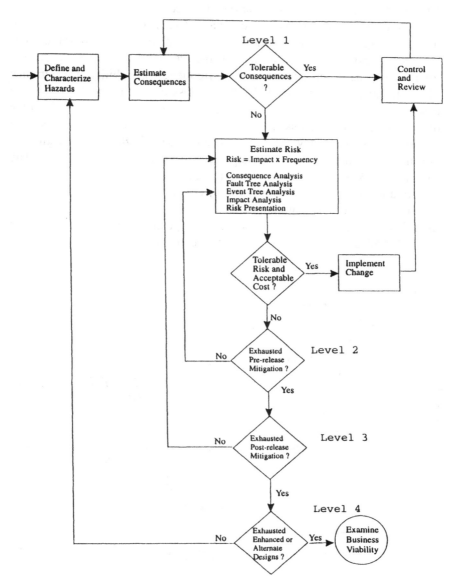

Figure 7.23. Risk-based approach selection and evaluation of mitigation systems
(G. A. Melhem and P. A. Croce, 1994).

Once the event/failure pathways for loss of containment scenarios have been identified, consequence analysis is used to establish the extent of damage and/or hazard footprints. For simple cases, many operating companies employ qualitative criteria to categorize the consequences of the identified failure modes. For complex cases, complete and detailed calculations are utilized to assess the hazard impact.

The estimated impact is then compared to hazard acceptance criteria to determine whether the consequences are tolerable without additional loss prevention and mitigation measures. If the identified consequences are not tolerable, the next step is to estimate the frequency/probability of occurrence of the identified failure modes leading to loss of containment. For simple cases, frequency estimates are combined with consequences to yield a qualitative estimate of risk. For complex cases, fault tree analysis is used to estimate the frequency of the event leading to the hazard. These estimates are then combined with the consequences to yield a measure of risk. The calculated risk level is compared to a risk acceptance criterion to determine if mitigation is required for further risk reduction.

If risk reduction is required, prerelease mitigation measures are typically used first. If the risk is still not acceptable, postrelease mitigation is then used. Postrelease mitigation is less cost-effective than prerelease mitigation.

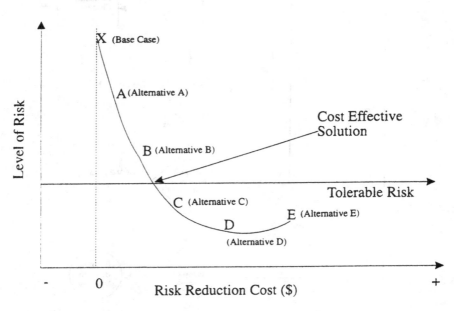

Figure 7.24. Impact of alternate designs or mitigation measures on risk reduction (G. A. Melhem and P. A. Croce, 1994).

If the risk level is still not acceptable, or if the cost of mitigation is excessive, enhanced or inherently safer design techniques are required. At this stage the entire cycle is repeated.

The outcome of the risk-based approach shown in Figure 7.23 is illustrated in Figure 7.24. The outcome can be either qualitative or quantitative. Figure 7.24 shows the impact of alternate designs or mitigation measure on risk reduction. A cost-effective solution is one where the risk is reduced to an acceptable level at a reasonable cost.

7.10. References

AIHA (American Industrial Hygiene Association). 1992. *Emergency Response Planning Guidelines for Air Contaminants.* Fairfax, VA: American Industrial Hygiene Association.

Blewitt D. N., R. L. Petersen, M.R. Ratcliff, and G. Heskestad. 1991. *Evaluation of Water Spray Mitigation System for an Industrial Facility.* Center for Chemical Process Safety, New York: American Institute of Chemical Engineers.

CCPS (Center for Chemical Process Safety). 1989. *Guidelines for Chemical Process Quantitative Risk Analysis.* New York: American Institute of Chemical Engineers.

CCPS (Center for Chemical Process Safety). 1996. *Guidelines for Use of Vapor Cloud Dispersion Models, Second Edition.* New York: American Institute of Chemical Engineers.

GPSA (Gas Processors Suppliers Association). 1972. *Engineering Data Book, 9th Edition.* Tulsa, OK: Gas Processors Suppliers Association.

Fthenakis, V. M., and D. N. Blewitt. 1992. Mitigation of Hydrofluoric Acid Releases: Simulation of the Performance of Water Spray Systems. *Journal of Loss Prevention in the Process Industries.* 6(4).

Fthenakis, V. M., and D. N. Blewitt. 1994. Recent Developments on Modeling Mitigation of Accidental Releases of Hazardous Gases. *Process Plant Symposium.* New York: American Institute of Chemical Engineers.

Fthenakis, V. M., K. W. Schatz, and V. Zakkay. 1990. *Proceedings of the International Conference and Workshop on Modeling and Mitigating the Consequences of Hazardous Materials.* New York: American Institute of Chemical Engineers.

Melhem, G. A., and P. A. Croce, 1994. "Advanced Consequence Analysis: Emission, Dispersion. Fires and Explosions Dynamics," Working Manuscript, Arthur D. Little, Inc.

Petersen, R. L., and D. N. Blewitt. 1992. *Evaluation of Water Sprays/Fire Monitor Mitigation for Two Refineries.* New York: American Institute of Chemical Engineers.

Post, L. 1994. "New Options for the HEGADAS Model of HGSYSTEM." Shell Research.

Schatz, K. W., and R. P. Koopman. 1989. "Effectiveness of Water Spray Mitigation Systems for Accidental Releases of Hydrogen Fluoride." Industrial Cooperative HF Mitigation Assessment Program Water Spray Subcommittee.

Witlox, H. W. M., K. McFarlane, M. R. Ratcliff, and G. Heskestad. 1991. *Development and Vo' d .tion of Atmospheric Dispersion Models for Ideal Gases and Hydrogen Fluoride.* Sh. International Research, Netherlands: Maatschappij B. V.

Index

Publications Available from the
CENTER FOR CHEMICAL PROCESS SAFETY
of the
AMERICAN INSTITUTE OF CHEMICAL ENGINEERS

CCPS Guidelines Series

Guidelines for Postrelease Mitigation in the Chemical Process Industry

Guidelines for Integrating Process Safety Management, Environment, Safety, Health, and Quality

Guidelines for Use of Vapor Cloud Dispersion Models, Second Edition

Guidelines for Evaluating Process Plant Buildings for External Explosions and Fires

Guidelines for Writing Effective Operations and Maintenance Procedures

Guidelines for Chemical Transportation Risk Analysis

Guidelines for Safe Storage and Handling of Reactive Materials

Guidelines for Technical Planning for On-Site Emergencies

Guidelines for Process Safety Documentation

Guidelines for Safe Process Operations and Maintenance

Guidelines for Process Safety Fundamantals in General Plant Operations

Guidelines for Chemical Reactivity Evaluation and Application to Process Design

Tools for Making Acute Risk Decisions with Chemical Process Safety Applications

Guidelines for Preventing Human Error in Process Safety

Guidelines for Evaluating the Characteristics of Vapor Cloud Explosions, Flash Fires, and BLEVEs

Guidelines for Implementing Process Safety Management Systems

Guidelines for Safe Automation of Chemical Processes

Guidelines for Engineering Design for Process Safety

Guidelines for Auditing Process Safety Management Systems

Guidelines for Investigating Chemical Process Incidents

Guidelines for Hazard Evaluation Procedures, Second Edition with Worked Examples

Plant Guidelines for Technical Management of Chemical Process Safety,
    Revised Edition
Guidelines for Technical Management of Chemical Process Safety
Guidelines for Chemical Process Quantitative Risk Analysis
Guidelines for Process Equipment Reliability Data, with Data Tables
Guidelines for Safe Storage and Handling of High Toxic Hazard Materials
Guidelines for Vapor Release Mitigation

CCPS Concepts Series

Inherently Safer Chemical Processes: A Life-Cycle Approach
Contractor and Client Relations to Assure Process Safety
Understanding Atmospheric Dispersion of Accidental Releases
Expert Systems in Process Safety
Concentration Fluctuations and Averaging Time in Vapor Clouds

Proceedings and Other Publications

Proceedings of the International Conference and Workshop on Process Safety
    Management and Inherently Safer Processes, 1996
Proceedings of the International Conference and Workshop on Modeling and
    Mitigating the Consequences of Accidental Releases of Hazardous
    Materials, 1995.
Proceedings of the International Symposium and Workshop on Safe Chemical
    Process Automation, 1994
Proceedings of the International Process Safety Management Conference and
    Workshop, 1993
Proceedings of the International Conference on Hazard Identification and Risk
    Analysis, Human Factors, and Human Reliability in Process Safety, 1992
Proceedings of the International Conference and Workshop on Modeling and
    Mitigating the Consequences of Accidental Releases of Hazardous
    Materials, 1991.
Safety, Health, and Loss Prevention in Chemical Processes: Problems for
    Undergraduate Engineering Curricula